Evolution Propelling
Through Communication

Dr Marius Albu

Foreword

Although communication is so much used in our daily life as the conveyor of information needed in almost all fields of activity or interest, we are still far to comprehend its essential role played during the evolution of species in general and of humans in particular.

On the entire scale of evolution from non-cellular organisms to humans, any form of life has specific senses, by which it is able to receive, transmit and send signals, both inside and outside its body. Notably, signalling together with self-sustaining processes make the animate matter distinguishable from the inanimate one.

Communication can not be conceived without life, whilst the life itself could not evolve without communication. Their interdependence is the key for understanding both the evolution of organisms from simplest to most complex species, and the development of communication from simple signalling to high technology communication. For instance, between about 1000 and 700 million years ago, some of the earlier grazing and burrowing small organisms increased in abundance, diversity, complexity of shape, and especially size and number of spines which made them *predators* and imposed upon their preys the adjustment of defensive means, thus impelling the development of communication that in turn played an important part in the outburst of higher forms of life.

Evolution of life propelling through development of communication was marked by successive stages with their representatives, characteristics and timelines, displayed as follows: life and forms of communication, primates and their communication, hominids and roots of speech, humans and language, writing and printing, literature and its extent, and finally communication technology. Altogether this long lasting process takes place according to the general constraints of conserving its total energy and limiting its efficiency, which result in a time function useful to distinguish the appropriate sequences for the entire course of evolution through communication.

Abbreviations and symbols

AD	*Anno Domini*	P	brain power
ANS	autonomic nervous system	p	brain potential; pulse
B_w	body weight	$\wp$	probability
b_w	brain weight	P_t	*t-power* of evolution
BC	Before Christ	P_τ	*τ-power* of evolution
BP	Before Present	PC	personal computer
C	coulomb	Q	quotient of efficiency
C	cephalisation factor	q	brain volume or capacity
c	light speed	r	exponential constant
c.	(Latin *circa*) about	RNA	ribonucleic acids
DNA	deoxyribonucleic acids	S	entropy
E	brain energy; total energy	s	second
E_t	*t-energy*	t	(running) time
E_τ	*τ-energy*	$t_o = 0$	initial time
e-book	electronic book	$t_\bullet$	final time
e.g.	(Latin *exempli gratia*) for example	*tanh*	hyperbolic tangent
email	electronic mail	*tanh⁻¹*	inverse hyperbolic tangent
EQ	encephalization quotient	V	volt
f	*(2π)·*(frequency)	W	watt
h	Planck's constant	www	World Wide Web
i	square root of (-1); number of order	α	first-order sequences
i.e.	(Latin *id est*) that is	β	second-order sequences
J	joule	γ	third-order sequences
K	kelvin	δ	fourth-order sequences
k	Boltzmann's constant	ε	fifth-order sequences
kg	kilogram	ζ	sixth-order sequences
log	logarithm	η	seventh-order sequences
ln	natural logarithm	θ	rotational angle
m	metre (meter)	λ	lifespan; wavelength
N	newton	ν	frequency
N	set of positive integers	ρ	density of brain energy
n	total number of bits	Σ	sum of terms
τ	period	$=$	equal to
$\Phi(v)$	Fourier transform	$\approx$	approximately equal to
$\varphi(t)$	function of time	$>$	greater than
ω	angular speed	$<$	less than
$+$	plus	$\geq$	greater than or equal to
-	minus	$\leq$	less than or equal to
\	separation	∞	infinity

±	plus or minus	∫	integral
·	multiply by	@	*at* sign
/	divide by	©	Copyright symbol

CONTENTS

1. Life and forms of communication

1.1. Origins of life and communication

The activity of conveying information through the exchange of signals, behaviour, visuals, messages, thoughts, speech, writing, mass media (newspaper, article, book, pamphlet, comics, radio, film, television), or digital media (internet, email, websites, blogs), and so on, as the meaningful exchange of information between two or more living creatures, is called _communication_ (from Latin _communico, communicare_ 'to share out, give a share in; to join, unite; to take a share, participate'). In general, the communication can be defined as an act by which one organism sends or receives from another warning or information about its needs, perceptions, affective states, desires, or knowledge, which may be unintentional or intentional, by unconventional or conventional signals, and in non-linguistic or linguistic forms. The study of signs and sign processes, as well as indication, designation, likeness, analogy, metaphor, symbolism, signification, and communication - as the science of signs or signals - is called _semiotics_ (Greek σημιωτική, from σημειον 'sign'), the term being introduced into English (J Locke, 1664-90) as a synonym for 'doctrine of signs' (Latin _doctrina signorum_ 'instruction of signs'). Subsequently, this study was divided into three main branches: _semantics_ (relationships between signs and symbols, and what they represent), _syntactics_ (formal properties of signs, symbols and languages), and _pragmatics_ (relations between signs or expressions and their users). _Semiotic biology_, or _biosemiotics_, has been pioneered (JJ von Uexkül, 1920-35), and a part of it was initiated (TA Sebeok, 1968-72) and developed (T Maran _et al._, 2011; G Witzany, 2014) as _zoosemiotics_, the semiotic study of the use of signals among the animals, i.e. the study of how something comes to function as a sign to an animal; which played an important part in the development of ethology, sociobiology, and animal cognition.

The investigation of meaning and representation of communication has been approached by a model of _signalling game_ designed for applying to communication between two human agents (D Lewis, 1969), treating them from an evolutionary point of view, which was later extended by _sender-receiver models_ applicable to the communication between nonhuman organisms, from bacteria to mammals (B Skyrms, 1995-2010). _Communication cycle_ (CE Shannon and W Weaver, 1948-49) consists of the following elements: _(i)_ an _information source_ that produces a message, *(ii)* a *transmitter* that encodes the message into signals, *(iii)* a *channel* to which signals are adapted for transmission, *(iv)* a *receiver* that 'decodes' (reconstructs) the message from signal, and *(v)* a *destination* where the message arrives. The study of communication continued with intra-species signalling (in bacteria, yeast, social insects, and also many vertebrates) and interspecies signalling (in bacteria, and also between the gut flora and their host, as part of their commensalism or symbiotic relationship).

Communication implies the existence of life

Without life, the communication is not conceivable. In its simplest form, communication consists in signalling, i.e. an elementary sender - signal - receiver system of a living creature. Taking into account that communication can be found where life would be, scientists are using new technological means (infrared emissions, spectral analyses, etc.) to identify the existence of life outside the Earth, focusing on spatial scales such as the Universe, the Milky Way, and the Solar System (M Albu, 2013-14). Some results of their works are mentioned below.

In the *Universe*, recently spectral analyses indicate that organic compounds are commonly present throughout it as a mixture of *aromatic* (ring-like) and *aliphatic* (chain-like) *structures*, comprising up to one hundred carbon atoms, and resembling those of coal and petroleum. Since coal and oil are considered fossil remnants of life, this type of organic matter is thought to arise from living organisms (S Kwok and Y Zhang, 2010-11). Such a discovery shows that: *(i)* complex organic compounds can be synthesized in space, even when no life forms are present; *(ii)* stars are making such compounds on extremely short time scales of weeks; *(iii)* old stars are capable of producing organic compounds; and, most interestingly, *(iv)* organic star dust is similar in structure to the complex organic compounds frequently found in meteorites. In 2012, astronomers at the Copenhagen University first reported the detection of a *glycolaldehyde* (sugar) molecule in a distant star system, identified around the proto-stellar binary *IRAS 16293-2422* at about 400 light-years from us; being known that such molecules can participate in formation of the ribonucleic acids (RNA). In the same year, NASA scientists reported that *polycyclic aromatic hydrocarbons*, subjected to interstellar medium conditions, are transformed, through hydrogenation, oxygenation and hydroxylation, to more complex organics and thus 'a step along the path toward amino acids and nucleotides, the raw materials of proteins and DNA respectively'. Spectroscopic investigations, using telescopes, detected interstellar *formaldehyde*, the first polyatomic organic molecule to be found in the interstellar medium. Meanwhile, over a hundred interstellar species were identified by radio astronomy; these including radicals and ions, carbon-based compounds such as *alcohols*, *acids*, *aldehydes*, and *ketones*, and also carbon monoxide as a common interstellar molecule.

In the <u>Milky Way</u>, two *complex organic molecules* were detected in its 'sweet spot', suggesting that the building blocks for life may have existed in this region even before the formation of planets. Recently, a giant molecular cloud of gas and dust, named *Sagittarius B2*, was discovered at about 390 light-years from the centre of our galaxy, which is something akin to a galactic watering hole; this cloud contains a wide variety of organics, including two of the most complex molecules ever found in interstellar space, namely *ethyl* and *n-propyl cyanide*, similar in size and complexity to amino acids, the basic components of organisms. In 2000-01, using the National Science Foundation's 12-Meter Telescope on Kitt Pick in Arizona, US scientists discovered, in a giant cloud of dust and gas near the centre of our galaxy, a simple sugar molecule *glycolaldehyde* that is an 8-atom molecule composed of carbon, oxygen and hydrogen, which can combine with other molecules to form more complex sugars such as *ribose* and *glucose*. The Atacama Large Millimetre Array (ALMA Project) confirmed in 2013 that its researchers have discovered an important pair of *pre-biotic molecules* in icy particles found in a giant cloud of gas at about 25000 light-years from us.

In the <u>Solar System</u>, recent studies have led to the conclusion that bacteria may travel dormant for an extended amount of time before colliding randomly with other planets or intermingling with proto-planetary disks. A 2008 analysis of the (carbon-12)/(carbon-13) ratio of *organic compounds* found in the *Murchinson meteorite* indicates a non-terrestrial origin for these molecules. Biologically relevant molecules identified so far include *uracil*, an RNA nucleobase, and *xanthine*. These results demonstrate that many organic compounds which are components of life on Earth have been already present in the young Solar System and may have played a key role in life's origin on our planet. In 2009, NASA scientists identified one of the fundamental chemical building-blocks of life, the amino acid *glycine*, in a comet for the first time. Other recent research showed that organics such as *amino acids* and *nucleobases* result from high-energy ultraviolet radiation bombarding simple ices (S Sanford, 2011-12).

Compared with other planets in our Solar System, Mars seems to have, at least in its early stage, favourable conditions for emergence of simple forms of life. Consequently, scientists were and still are hoping to discover there some remains from an incipient form of life. Starting in 1977, two Viking Landers scooped up Martian soil, heated it, and analyzed the gases that came off; in those samples they detected *chloromethanes*. Subsequently, in 2012, NASA began the *Curiosity mission* for probing Mars in order to find out whether or no life having existed on the Red Planet. An analysis of a rock sample collected by NASA's Curiosity rover shows that ancient Mars could have supported living microbes. Indeed, the first full analysis of Martian soil by this rover has detected simple carbon compounds that could be the first traces of past Martian life ever found. In 2013, by this Curiosity rover, there were detected trace amount of three of the simplest possible carbon-containing compounds: a carbon atom with one, two, or three chlorine atoms attached in place of hydrogen atoms. It will now take more samples to be analyzed in more ways, and compared to some carbon-free blank samples brought from Earth. Meanwhile, *methane plumes* were discovered in the Mars' atmosphere, a tantalizing clue pointing to potential subsurface life and raising the hopes of some experts. Alien bugs may be responsible for these plumes of methane gas, but the gas emissions could have either a geological or biological source. After studying the first measurements of the Martian atmosphere taken by the Curiosity rover, researchers concluded that the Red Planet lost its oxygen four billion years ago, possibly due to a massive collision with a spatial body. Some scientists had already declared that the 'seeds' of life probably arrived on the Earth in meteorites blasted off Mars by impacts or volcanic eruptions, believing that the element molybdenum was a catalyst that helped organic molecules develop into the first living things. Further, they argued that this form of molybdenum could not have been available on Earth at the time life first began, because until about three billions years ago the surface of the Earth had very little oxygen, but Mars did (S Benner, 2013). NASA's Curiosity rover also found evidence that conditions on Mars were once suitable for microbial life after analyzing the first drilled sample on Martian rock, 'John Klein' rock, at Yellowknife Bay in Gale Crater, where it was detected water, carbon dioxide, oxygen, sulphur dioxide, hydrogen sulphide, chloromethane and dichloromethane. Meanwhile, analysis of a Martian meteorite recently showed that there was also boron on Mars, and that the oxidized form of molybdenum was there too. Another reason why life would have struggled to start on early Earth was that it was likely to have been covered by water. After discovering water on the Moon, other scientists are intending to put a satellite in orbit around Mars to find out more about the traces of life in the early stage of this planet.

On the _Earth_, the life appeared for once when the first living molecules occurred, but to be formed, survive and proliferate they needed an environment with water and maybe some oxygen, which also had to be present there.

Initially, the 4568 million year old Earth had no water and the first atmosphere, consisting mainly of hydrogen and helium, had been driven off by the solar wind. Water was later supplied by icy meteorites from the outer asteroid belt and some large planetary embryos, as well as by comets, altogether heavily bombarding the young Earth and carrying meteoritic water and water vapour that condensed, and thus the first accumulations of water formed on the Earth's surface. The existence of the first ocean was proved by identification of _detrital grains of zircon_ in rocks of the _Nuwagittuq Greenstone Belt_ (Canada), _Isua Greenstone Belt_ (Southwest Greenland), _Narryer Gneiss Terrane_ (Western Australia), and _Acasta Gneiss_ in the Canadian Shield (Northeast Territories, Canada), which were radiometrically dated (on their content in uranium and thorium) to more than 4400 million years ago and indicated contact with liquid water (S Bowring and I Williams, 1999; S Wilde _et al._, 2000-01; C Fedo _et al._, 2001). The same age has been confirmed by the oxygen isotopic compositions of some of these zircons, showing that at that time there was already water on the Earth's surface, and also at least traces of oxygen. Meanwhile, volcanic mater brought to the surface ejected steam, carbon dioxide and ammonia, forming a second atmosphere consisting largely of carbon dioxide and water vapour together with nitrogen and inert gases, which was suitable for starting the early carbon cycle. In the primeval ocean on the Earth's surface, the first organisms appeared, either from extraterrestrial simple forms of life pre-existing in the universe and carried to the Earth by the bodies bombarding our planet (D Raoult _et al._, 1992-2004; B La Scola _et al._, 2003), or from terrestrial inorganic matter present in the early ocean that spontaneously proceeded into organic molecules and then cells (HC Urey, 1945-61; SL Miller, 1953-60).

The earliest known traces of life have been dated back 4000 million years, according to the _biochemical banded iron formations_ formed in sea water as a result of oxygen released by photosynthetic cyanobacteria (blue-green algae), which are found in many places, including Isua in West Greenland, with ages of 3800-4000 million years (A Lepland _et al._, 2002; A Mloszewska and K Konhauser, 2005-

12

10). As the presence of detrital zircon crystals indicates that water was present on the Earth's surface about 4400 million years ago, and the existence of photosynthetic cyanobacteria consuming carbon dioxide and releasing oxygen is proved by analyses of banded iron formations with ages up to 4000 million years, it follows that the first forms of life originated around 4200 million years ago and together with them the *first communication* on the Earth.

Efficient communication requires tight co-evolution between the signal emitted and the response elicited. The evolutionary conditions for emerging communication have been studied by experiments with *robots*, and showed that communication readily evolves when colonies consist of genetically similar individuals and when selection acts at the colony level. By these experiments, there were identified several distinct communication systems differing in their efficiency, but, once a given system of communication was well established, it constrained the formation of more efficient communication systems. Under individual selection, the ability to produce signals resulted in the development of deceptive communication strategies in colonies and a concomitant decrease in colony performance. However, there are predictions about the evolutionary conditions conducive to the emergence of communication and provides guidelines for designing artificial evolutionary systems displaying spontaneous communication (S Mitri *et al.*, 2008).

1.2. Characteristics of various organisms

On the entire scale of evolution, the organisms have specific kinds of communication, depending on the taxonomic rank to which they are belonging: from domain, kingdom and phylum; through class, order and family; to genus and species. These kinds of communication are intrinsically controlled and co-ordinated by two fundamental processes, namely the genetic heritage and the neural signalling.

Genetic heritage

Any organism has its units of heredity called <u>*genes*</u>, each of them consisting of stretches of *ribonucleic acids* (RNA), i.e. nucleic acids containing ribose (a pentose as $C_5H_{10}O_5$), present in living cells and responsible for the development of proteins; and *deoxyribonucleic acids* (DNA), i.e. nucleic acids containing deoxyribose (a deoxy-sugar as $C_5H_{10}O_4$), present in chromosomes and consisting of complex molecules.

Each *chromosome* is a structure of DNA, protein, and RNA present in a cell, that represents a single piece of coiled DNA containing many genes, regulatory elements and other nucleotide sequences; within a cell-nucleus, it looks as a rod-like portion of chromatin (a readily stained substance in the nucleus of a cell) performing an important part in mitotic cell-division and in heritable transmission in all plant and animal cells, which carries in coded form instructions for passing on hereditary characteristics (W Johannsen, 1909; GW Beadle and EL Tatum, 1942; OT Avery and M McCarty, 1944; S Ochoa, 1948-67; F Jacob *et al.*, 1950-65; EL Tatum and J Lederberg, 1950-75).

One major difference between RNA and DNA is the kind of sugar contained, as *pentose sugar ribose* in RNA, and as *2-deoxyribose* in DNA. Usually DNA occurs as circular chromosomes in prokaryotes, and as linear chromosomes in eukaryotes.

Living creatures depend on genes, as they specify all proteins and functional RNA chains. Genes hold the information to build and maintain an organism's cells and pass genetic traits to offspring.

When proteins are manufactured, the gene is first copied into RNA as an intermediate product. In other cases, the RNA molecules are the actual functional products. For example, RNAs known as *ribozymes* are capable of enzymatic function, and microRNA has a regulatory role. DNA sequences from which such RNAs are transcribed are known as *RNA genes*.

There are two types of cells in a body: *haploid cells*, containing only one set of chromosomes (half of the number of chromosomes in somatic cells); and *diploid cells*, containing two complementary sets of chromosomes.

Some viruses store their entire genomes in the form of RNA and contain no DNA at all. RNA is assembled as a typically *single-stranded* chain of nucleotides, each of them containing a *ribose sugar*, to which is attached one of the bases called *adenine* (*A*), *cytosine* (*C*), *guanine* (*G*), or *uracil* (*U*); where, notably, the complementary base to adenine is uracil (an unmethylated form of thymine).

The vast majority of living organisms encode their genes in strands of the polymer DNA with a commonly *double-stranded* structure, i.e. comprising *two helical chains*, coiled round the same axis and twisted around each other in a right-handed spiral (FHC Crick and JD Watson, 1953). Each DNA strand, or chain, is a monomer unit, called nucleotide, that consists of a *five-carbon sugar* (2'-deoxyribose), a *nitrogenous base* attached to sugar, and a *phosphate group*. The component nucleotides can be of four types differing from each other only by the nitrogenous base, as one of the four possible bases called *adenine* (*A*), *cytosine* (*C*), *guanine* (*G*) and *thymine* (*T*); where *A* and *G* are purines ($C_5H_4N_4$), while *C* and *T* are pyrimidines ($C_4H_4N_2$).

Genes that encode proteins are composed of a series of three-nucleotide sequences called *codons*, which serve as the 'words' in the 'genetic language'. Genetic code specifies the correspondence during protein translation between codons and amino acids, being nearly the same for all known organisms.

Many prokaryotic genes are organized into *operons*, or group of genes whose products have related functions and which are transcribed as a unit. By contrast, eukaryotic genes are transcribed only one at a time, but may include long stretches of DNA called *introns* which are transcribed but never translated into protein, being spliced out before translation. Splicing can also occur in prokaryotic genes, but is less common than in eukaryotes.

The set of chromosomes in a cell makes up its <u>genome</u>, the full set of chromosomes of an individual (H Winker, 1920). Large-scale genome evolution indicates a close coordination between the sequence and size of genomes. The genome is encoded either in DNA or, for many types of virus, in RNA, and includes both the genes and the non-coding sequences of DNA/RNA.

DNA is also present in the *mitochondria*, which are energy-producing bodies, thread-like to spherical in shape, present in cytoplasm, generating about 90% of the energy needed for the body's function.

15

In order to exemplify the genetic background of organisms in their evolution, the genome sizes of some species are briefly presented below with their estimated numbers of base pairs and protein-encoding genes (L Prey, 2008).

Species (common name)	Base pairs (millions)	Protein-encoding genes (thousands)
Saccharomyces cerevisiae (unicellular budding yeast)	12	6
Trichomonas vaginalis (anaerobic flagellated protozoan)	160	60
Plasmodium falciparum (unicellular malaria parasite)	23	5
Caenorhabditis elegans (nematode)	95.5	18
Drosophila melanogaster (fruit fly)	170	14
Arabidopsis thaliana (mustard; thale cress)	125	25
Oryza sativa (rice)	470	51
Gallus gallus (chicken)	1000	20...23
Canis familiaris (domestic dog)	2400	19
Mus musculus (laboratory mouse)	2500	30
Homo sapiens (human)	2900	20...25

The region of the chromosome at which a particular gene is located is called its *locus*. A chromosome consists of a single, very long DNA helix on which thousands of genes are encoded. Prokaryotes, such as bacteria and archaea, typically store their genomes on a single large, circular chromosome, sometimes supplemented by additional small circles of DNA, called *plasmids*, and can be passed between individual cells, even those of different species, via horizontal gene transfer. Although some simple eukaryotes also possess plasmids with a small number of genes, the majority of eukaryotic genes are stored on multiple linear chromosomes, which are packed within the nucleus with storage proteins called *histones*.

Simple single-celled eukaryotes have relatively small amounts of so-called 'junk DNA', whereas the genomes of complex multicellular organisms, including humans, contain a vast majority of DNA without an identified function. Although protein-coding DNA makes up barely 2% of the human genome, about 80% of the bases in the genome may be expressed, so the term 'junk DNA' may be a misnomer.

In all organisms, there are two major steps separating a protein-coding gene from its protein: 1st, the DNA on which the gene resides must be transcribed from DNA to a messenger RNA (mRNA); and 2nd, it must be translated from mRNA or protein. RNA-coding genes must still go through the first step, but are not translated into protein.

The process of producing a biologically functional molecule of either RNA or protein is called *gene expression*, and the resulting molecule itself is called a *gene product*.

The *genetic code* is the set of chemical symbols by which a gene is translated into a functional protein. Each gene consists of a specific sequence of nucleotides encoded in a DNA (or RNA in some viruses) strand; a correspondence between nucleotides, the basic building blocks of genetic material, and amino acids, the basic building blocks of proteins, must be established for genes to be successfully translated into functional proteins. Sets of three nucleotides, known as codons, each corresponds to a specific amino acid or to a signal; three codons are known as 'stop codons' and, instead of specifying a new amino acid, alert the translation machinery when the end of the gene has been reached. There are 64 possible codons (four possible nucleotides at each of three positions hence 4^3 possible codons) and only 20 strand amino acids; hence the code is redundant and multiple codons can specify the same amino acid. The correspondence between codons and amino acids is nearly universal among all known living organisms.

The process of genetic *transcription* produces a single-stranded RNA molecule known as *messenger RNA*, whose nucleotide sequence is complementary to the DNA from which it was transcribed (GE Palade, 1956; D Baltimore, 1970; RJ Britten and EH Davidson, 1981-99; RD Kornberg, 1990-2005). The DNA strand whose sequence matches that of the RNA is known as the *coding strand*, and the strand from which the RNA was synthesized is the *template strand*.

In eukaryotes, transcription necessarily occurs in the nucleus, where the cell's DNA is sequestered; the RNA molecule produced by polymerase is known as *primary transcript* and must undergo *post-transcriptional modifications* before being exported to the cytoplasm for translation. The *splicing of introns* present within the transcribed region is a modification unique to eukaryotes; *alternative splicing* mechanisms can result in mature transcripts from the same gene having different sequences and thus coding different proteins. This is a major form of regulation in eukaryotic cells.

Translation is the process by which a mature mRNA molecule is used

17

as a template for synthesizing protein. It is carried out by *ribosomes*, large complexes of RNA and protein responsible for the chemical reactions to add new amino acids to a growing *polypeptide chain* by the formation of *peptide bonds*. The genetic code is read three nucleotides at a time, in units called codons, via interactions with specialized RNA molecules called transfer RNA (tRNA). Each tRNA has three unpaired bases known as *anticodon*, i.e. complementary to the codon it reads; the tRNA is also covalently attached to the amino acid specified by the complementary codon. When the tRNA binds to its complementary codon in an mRNA strand, the ribosome ligates its amino acid cargo to the new polypeptide chain, which is synthesized from *amino terminus* to *carboxyl terminus*. During and after its synthesis, the new protein must fold to its active *three-dimensional structure* before it can carry out its cellular function.

The growth, development and reproduction of organisms are based on *cell division* as the process by which a single cell divides into two usually identical *daughter cells*. This requires first making a duplicate copy of every gene in the genome, in a process called *DNA replication*. The copies are made by specialized enzymes known as *DNA polymerases*, which 'read' one strand of the double-helical DNA, known as the *template strand*, and synthesize a new complementary strand (A Kornberg, 1959-67). Because the DNA double helix is held together by *base pairing*, the sequence of one strand completely specifies the sequence of its complement; hence only one strand needs to be read by the enzyme to produce a faithful copy. The process of DNA replication is semi-conservative, so that the copy of the genome inherited by each daughter cell contains one original and one newly synthesized strand of DNA. After DNA replication is complete, the cell must physically separate the two copies of the genome and divide into two distinct membrane-bound cells.

In prokaryotes, such as bacteria and archaea, this usually occurs via a relatively simple process called *binary fission*, in which each circular genome attaches to the *cell membrane* and is separated into the daughter cells as the membrane invaginates to split the cytoplasm into two membrane-bound portions. Binary fission is extremely fast compared to the rates of cell division in eukaryotes.

Eukaryotic cell division is a more complex process known as the *cell cycle*, in which DNA replication occurs during a phase of this cycle known as the *S phase*, whereas the process of segregating chromosomes and splitting the cytoplasm occurs during the *M phase*. In many single-celled eukaryotes, such as yeast, reproduction by *budding* is common, which results in asymmetrical portions of cytoplasm in the

18

two daughter cells.

The duplication and transmission of genetic material from one generation of cells to the next is the basis for _molecular inheritance_, and the link between the classical and molecular pictures of genes. Organisms inherit the characteristics of their parents because the cells of the offspring contain copies of the genes in their parents' cells (GJ Mendel, 1856-67; CR Darwin, 1862-71; AFL Weismann, 1867-1902; M Delbrück, 1940-46). In _asexually reproduction_, the offspring will be a genetic copy or _clone_ of the parent organism. In _sexually reproduction_, a specialized form of cell division called _meiosis_ produces cells called _gametes_ or _germ cells_ that are haploid, i.e. contain only one copy of each gene. The gametes produced by females are called _eggs_ or _ova_, and those produced by males are called _sperm_. Two gametes fuse to form a _fertilized egg_, a single cell that once again has a _diploid_ number of genes, each with one copy from the mother and one copy from the father.

During the process of meiotic cell division, an event called _genetic recombination_ or crossing-over can sometimes occur, in which a length of DNA on one _chromatid_ is swapped with a length of DNA on the corresponding sister chromatid. The closer two genes lie on the same chromosome, the more closely they will be associated in gametes and the more often they will appear together; genes that are very close are essentially never separated because it is extremely unlikely that a cross-over point will occur between them. This is known as _genetic linkage_ (W Bateson, 1905-06; AH Sturtevant, 1911-13).

As a discipline of biology, _genetics_ is the science of genes, heredity and variation in living organisms, showing how differences between individuals are transmitted from one generation to the next, and how the information in the genes is used in the development and functioning of the adult organisms. Genetics was based on thorough research of the structure and role of DNA (AD Hershey and M Chase, 1952); double-helical representation of DNA (FHC Crick and JD Watson, 1953); base sequence of RNA (WH Stein and S Moore, 1954-58); genetic map and specific genes on DNA, deciphering of full code, nucleotide sequence and alteration of DNA (MW Nirenberg _et al._, 1966); protein-free precursor RNA, split genes and mapping technique for DNA analysis (T Cech, 1977; PA Sharp, 1977); altering of the genetic code and copying DNA (KB Mullis, 1983); gene sequence with mutations and oncogenes (M Bishop, 1985-89); as well as interference-gene silencing by double-stranded RNA and reprogrammed cells.

Giving that genes are universal to living organisms, genetics is applied to the study of all living systems, from viruses and bacteria, through plants and animals, to humans.

On the genetic scale, from the simplest to the most complex forms of life, the total *number of base pairs* varies between several and tens of thousands, while the *length of DNA* ranges between $2.3 \cdot 10^{-9}$ and $5.6 \cdot 10^{-3}$ metres. The formation of DNA requires an *extensive force* greater than or equal to $6 \cdot 10^{-11}$ newtons, a *positive torque* greater than $3 \cdot 10^{-20}$ newton-metres, and supports a *loading force* of $1.7 \cdot 10^{-7}$ newtons.

In eukaryotic organisms, during the meiotic metaphase I, the independent assortment produces a *gamete* with a mixture of organism's chromosomes. The physical basis of the independent assortment of chromosomes is the random orientation of each bivalent chromosome along the metaphase plate with respect to the other bivalent chromosome. Along with 'crossing over', this kind of assortment increases genetic diversity by producing novel genetic combinations.

Genetic development can be described by the DNA linearly extending in *length* from $2.3 \cdot 10^{-9}$ to $5.6 \cdot 10^{-3}$ metres, in *diameter* from $1.8 \cdot 10^{-9}$ to $2.3 \cdot 10^{-9}$ metres, in *cross-sectional area* from $2.54 \cdot \cdot 10^{-18}$ to $4.15 \cdot 10^{-18}$ square metres, in *volume* from $5.84 \cdot 10^{-27}$ to $2.32 \cdot 10^{-20}$ cubic metres, and in *time* from $4.2 \cdot 10^{-5}$ to $2.5 \cdot 10^{8}$ seconds. Therefore, the *speed of development* varies approximately from $5.5 \cdot 10^{-5}$ to $2.3 \cdot 10^{-11}$ metres per second. As the DNA force ranges from $6.0 \cdot 10^{-11}$ to $1.7 \cdot 10^{-7}$ newtons, it follows that the corresponding *energy* varies from $1.4 \cdot 10^{-19}$ to $9.5 \cdot 10^{-10}$ joules, and *power* varies from $3.3 \cdot 10^{-15}$ to $3.8 \cdot 10^{-18}$ watts (M Albu, 2013-14).

Heredity is defined as the passing of traits to offspring from its parents or ancestors. By this process an offspring cell or organism acquires or becomes predisposed to the characteristics of its parent cell or organism. Through heredity, variations exhibited by individuals can accumulate and cause some species to evolve.

Recent findings have confirmed important examples of inheritable changes that cannot be explained by direct agency of the DNA molecule. Such phenomena are classed as *epigenetic inheritance* systems that are causally or independently evolving over genes. This developing field of research into modes and mechanisms of epigenetic inheritance broadens the scope of heritability and evolutionary biology in general. Specifically, the *three dimensional conformation* of proteins (such as *prions*), *evolution in four dimensions* (E Jablonka and M Lamb, 2005), *transgenerational epigenetic inheritance*, *DNA methyllation* marking chromatin, self-sustaining *metabolic loops*, and gene silencing by *RNA interference* are areas where epigenetic inheritance systems have been discovered at the organismic level.

Heritability may also occur at even larger scales; for example, ecological inheritance through the process of *niche construction*, defined by the regular and repeated activities of organisms in their environment, leading to a legacy of effect that modifies and feeds back into the selection regime of subsequent generations; thus descendants inherit genes plus environmental characteristics generated by the ecological actions of ancestors. Other examples of heritability in evolution that are not under direct control of genes include the *morphogenesis by symbiogenesis, inheritance of cultural traits, group heritability,* and joint *multilevel and genetic effects* on response to genetic selection.

Genetic development during the life evolution follows a course marked by five phases: 1st, formation of *simple organic RNA and DNA molecules,* the genetic material for life, as simple chains of nucleotides; 2nd, appearance of *replicating RNA molecules* copying themselves and producing chains of nucleotides; 3rd, enclosure of *replicating molecules within a cell membrane*; 4th, formation of *cells with metabolic processes* out-competing those with older forms of metabolism; and 5th, evolvement of *first lineage of multicellular organisms.* During this development, an important role is played by the *mitochondrial DNA* with its unique pattern of inheritance, passing down directly from mother to child, and accumulating changes much slower than other types of DNA (R Raff, 1996; C Nüsslein-Volhard, 2006).

Neural signalling
Animal communication by nerves and nervous system in invertebrates, brain in vertebrates, and cerebral cortex in mammals and primates during their evolution are controlled and co-ordinated through neural signalling.

Action potentials are neural precursors, which evolved in single-celled eukaryotes, being necessary for their incipient neural activity. These potentials are based on calcium rather than sodium, and subsequently became adapted into neural electrical signalling in multicellular animals. In some colonial eukaryotes, such as *Obelia,* electrical signals propagate not only through neural nets, but also through epithelial cells in the shared digestive system of the colony.

Some organisms, such as sponges (*Porifera phylum*), have no cells connected to each other by synaptic junctions, i.e. they have no neurons, and therefore no nervous system. However, these organisms have *homologues of many genes* for a synaptic-like function, or more precisely a group of proteins that cluster together to form a structure resembling a postsynaptic density, as the signal-receiving part of a synapse. They allow the communication of one organism to another via calcium waves and other impulses, mediating some simple actions including whole-body contraction.

More evolved organisms, such as jelly fish, comb jellies and related animals, have *diffuse nerve nets* rather than a central nervous system, the nerve net being spread more or less evenly across the body, and in comb jellies concentrated near the mouth. These nerve nets consist of *sensory neurons* that pick up chemical, tactile, and visual signals; *motor neurons* that can activate contractions of the body wall; and *intermediate neurons* that detect patterns of activity in the sensory neurons and send signals to groups of motor neurons as a result. In some cases, the groups of motor neurons are clustered into *discrete ganglia*.

Organisms such as radiata, possess only two primordial cell layers, called *endoderm* and *ectoderm*, and have a relatively *unstructured nervous system* through their ectoderm, where neurons are generated from a special set of ectodermal precursor cells, which also serve as precursors for every other ectodermal cell type.

Nervous system in invertebrates was studied and explained starting from the 19th century and continuing until the present (AE Verrill and SI Smith, 1874; CH Turner, 1900-20; JY Cousteau, 1973-74; J Stevely and D Sweat, 2009).
Invertebrates include arthropods, molluscs, and numerous types of worms. Two groups of invertebrates have notably *complex brains*: the *arthropods* (insects, crustaceans, arachnids, and others), which have a central brain with three divisions and large optical lobes behind each eye for visual processing; and the *cephalopods* (octopuses, squids, and similar molluscs), among them the octopuses and squids having the largest brains of any invertebrates. Worms are the simplest organisms to have a *central nervous system* with an *anterior brain* connected to a nerve cord, thus prefiguring the central nervous systems of all more evolved organisms.
Since 600-550 million years ago when bilaterians (animals with left and right sides that are approximate mirror images of each other) emerged from a common ancestral worm-like organism, with their bodies shaped as a tube with a hollow gut cavity running from mouth to anus, the *nerve cord* was formed with an enlarged *ganglion for each body segment*, and an especially *large ganglion* 'brain' at the front.

Among their descendants, even mammals, including humans, have the segmented bilaterian body at the level of the nervous system. The *spinal cord* contains a series of segmental ganglia, each with *motor and sensory nerves* that innervate a portion of the body surface and underlying musculature. On the limbs, the layout of the innervation pattern is complex, but on the trunk it gives rise to a series of narrow bands. The top three segments belong to the *brain*, which consists of *forebrain, midbrain,* and *hindbrain.* Based on stages in their early embryonic development, the bilaterians are divided into two groups (superphyla) called protostomes and deuterostomes. The *protostomes* are the more diverse group, including arthropods, molluscs, and numerous types of worms, which possess a *nerve cord on the ventral side* of the body. The *deuterostomes* include vertebrates, as well as echinoderms, hemichordates (mainly acorn worms), and xenoturbellidans, all having the *nerve cord on the dorsal side* of the body.

The gastropods, cephalopods, and bivalves appeared 540-485 million years ago, and among them the molluscs have two *pairs of main nerve cords* (three in bivalves), the *visceral cords* serving the internal organs, and the *pedal cords* serving the foot. Both pairs run below the level of the gut, and include *ganglia* as local control centres in important parts of the body. Most pairs of corresponding ganglia on both sides of the body are linked by 'commissures' (relatively large bundles of nerves). The only ganglia above the gut are the *cerebral ganglia*, which sit above the oesophagus (gullet) and handle 'messages' from and to the eyes. The *pedal ganglia*, which control the foot, are just below the oesophagus and their commissure and connections to the cerebral ganglia encircle the oesophagus in a nerve ring.

The basic structure of the <u>*bilaterian nervous system*</u> can be directly observed in the simplest bilaterian worms, for example, the earthworms, which have dual nerve cords running along the length of the body and merging at the tail and the mouth. These *nerve cords* are connected by *transverse nerves*, like the rungs of a ladder, coordinating the two sides of the animal. Two ganglia at the head end function similarly to a *simple brain*, and photoreceptors on the animal's eyespots provide sensory information on light and dark. In the very small roundworm *Caenorhabditis elegans*, the nervous system is sexually dimorphic; the nervous systems of the two sexes, males and hermaphrodites, having different numbers of neurons and groups of neurons that perform sex-specific function: 383 neurons in males, and 302 neurons in hermaphrodites.

23

Along with insects and crustaceans, the arthropods have a nervous system made up of a series of *ganglia*, connected by a *ventral nerve cord* made up of two parallel connectives running along the length of the belly. Many arthropods have well-developed sensory organs, including compound eyes for vision and antennae for olfaction and pheromone sensation, the sensory information from these organs being processed by the <u>brain</u>. The head segment has one ganglion on each side, though some ganglia are fused to form a brain, also known as *supraesophageal ganglion*, and other large ganglia. In insects, so-called *giant fibre systems* allow rapid conduction of nerve impulses connecting parts of the brain to specific muscles in the legs or wings, their brain being anatomically divided into three specialized segments: *protocerebrum*, *deutocerebrum*, and *tritocerebrum*, which control a remarkable variety of behaviours for locomotion, obtaining food, mating, and aiding the survival of their offspring. Immediately behind the brain, each insect possesses a *subesophageal ganglion*, consisting of three pairs of fused ganglia, which control the mouthparts, the salivary glands and certain muscles.

<u>Brain in vertebrates</u> shares a common underlying form, which is most clearly revealed during the early stages of embryonic development. In its earliest form, the brain appears as three swellings at the front end of the neural tube; these swellings eventually becoming the *forebrain*, *midbrain*, and *hindbrain*, as prosencephalon, mesencephalon, and rhombencephalon respectively. In the earliest stages of brain development, these three parts are roughly equal in size. In many classes of vertebrates, such as fish and amphibians, the three parts remain similar in size in the adult, but in mammals the forebrain becomes much larger than the other parts, and the midbrain becomes very small. Neuroanatomists usually divide the vertebrate brain into six main regions: *telencephalon* (cerebral hemispheres), *diencephalon* (thalamus and hypothalamus), *mesencephalon* (midbrain), *cerebellum*, *pons*, and *medulla oblongata*. The mammals have three major components to the brain, namely the *neocerebellum*, added to the *cerebellum*, and the *neocortex* grown out of the front of the forebrain.

<u>Fishes</u> have quite small brains relative to body size, compared with other vertebrates, typically one-fifteenth the brain mass of a similarly sized bird or mammal. However, some fishes have relatively large brains, most notably *mormyrids* 'elephant fish' and *sharks*, which possess brains about as massive relative to body weight as birds and marsupials. The fish brain consists of several components. At the front are the *olfactory lobes*, a pair of structures that receive and process

24

signals from the nostrils via the two olfactory nerves. Behind the olfactory lobes is the two-lobed *telencephalon*, structurally equivalent to the cerebrum in higher vertebrates, which is involved mostly in olfaction. Together these structures form the *forebrain*.

Connecting the forebrain to the *midbrain* is the *diencephalon*, which performs functions associated with hormones and homeostasis. Just above the diencephalon lies the *pineal body*, which detects light, maintains circadian rhythms, and controls colour changes. The *midbrain*, or *mesencephalon*, is particularly involved in swimming and balance.

The *cerebellum* is a single-lobed structure, typically the biggest part of the brain, in some fishes being involved in *electrical sense*. The *brain stem*, or *myelencephalon*, is the brain posterior, and controls some muscles and body organs, in some fishes governing respiration and osmoregulation. Fish behaviour reveals that they possess *spatial memory*, and *visual discrimination*.

In *amphibians*, the nervous system is basically the same as in other vertebrates, comprising a central brain, a spinal cord, and nerves throughout the body. Their brain is less well developed than that of reptiles, birds, or mammals, but similar in morphology and function to that of fishes. The brain consists of equal parts *cerebrum*, *midbrain*, and *cerebellum*. The cerebrum processes sensory input, such as smell in the olfactory lobe and sight in the optic lobe, and altogether is the centre of behaviour and learning. The cerebellum is the centre of muscular coordination, the *medulla oblongata* controls some organ functions including heartbeat and respiration, and the *pineal body* is involved in hibernation and aestivation. Some amphibians, such as *caecilians*, possess *electroreceptors* to locate objects around them when submerged in water.

The *reptilian nervous system* contains the same basic part of the amphibian brain, but the reptile *cerebrum* and *cerebellum* are slightly larger, and possess twelve pairs of cranial nerves. Due to their short cochlea, the reptiles use *electrical tuning* to expand their range of audible frequencies. Among them, larger *lizards*, like the monitors, are known to exhibit *complex behaviour*, including *cooperation*.

Birds have two sexes, male and female, their sex being determined by the Z and W chromosomes, rather than by the X and Y chromosomes in mammals. The avian nervous system consists of a central nervous system, including the *brain* and *spinal cord*; and a peripheral nervous system, including the *cranial* and *spinal nerves*, *autonomic nerves* and *ganglia*, as well as *sense organs*.

25

The functions of the avian nervous system are to obtain (via sensory receptors) information about the internal and external environment; to analyse and respond to that information; to store information as *memory* and *learning*; and to coordinate outgoing motor impulses to skeletal muscles and the viscera (smooth muscle, cardiac muscle, and glands). The avian brain includes the *medulla, optic lobe, cerebellum,* and *cerebrum* (two cerebral hemispheres plus olfactory lobes). Comparing with the brains of reptiles, birds have relatively larger cerebral hemispheres and cerebella, and in addition have larger optic lobes and smaller olfactory bulbs.

Generally, in vertebrates, especially in mammals, there are three main components of the central nervous system: white matter, grey matter, and substantia nigra (part of the grey matter). The size of brain varies between less than one gram in the North American ruby-throated hummingbird and six kilograms in the blue whale. Intermediary sized-brains can be exemplified as 0.003 kilograms in rat, 0.2 kilograms in lion, and 1.3 - 1.4 kilograms in human.

The structure, function and disorder of vertebrate brain were subjects of study for many anatomists, physicians, cytologists, physiologists, histologists, psychiatrists, biochemists, pharmacologists, biologists, and neurologists (A Vesalius, 1543; W Cullen, 1778-79; C Golgi, 1873-1918; IP Pavlov, 1874-1936; SR y Cajal, 1892-1914; R Guillemin, 1970-77; H Jerison, 1973; R Holloway and D Post, 1982; P Greengard *et al.*, 1989-2000; G Shepherd, 1994; RE Hoffman *et al.*, 2004; S Grillner, *et al.*, 2005).

White matter consists mostly of glial cells and myelinated axons that transmit signals from one region of the cerebrum to another, and between the cerebrum and lower brain centres. This matter actively affects how the brain learns and functions, and meanwhile modulates the distribution of action potentials, as a relay and coordinating communication between different brain regions. The main functions of white matter are: *projection tracts*, which extend vertically between the higher and lower brain, and *spinal cord centres*, carrying information between the cerebrum and the rest of the body, as well as signals upward to the cerebral cortex; *commissural tracts*, which cross from one cerebral hemisphere to the other through bridges called commissures, enabling the left and right sides of the cerebrum to communicate with each other; and *association tracts*, which connect different regions within the same hemisphere of the brain, long association fibres connecting different lobes of a hemisphere to each other, whereas short association fibres connect different gyri within a

26

single lobe; one of their roles is to link the perceptual and memory centres of the brain. White matter forms the bulk of the deep parts of the brain and the superficial parts of the spinal cord. The deep cerebellar white matter is called 'arbor vitae'.

Grey matter is a major component of the central nervous system consisting of *neuronal cell bodies*, *neuropil* (dendrites and unmyelinated axons), *glial cells* (astroglia and oligodendrocytes), and *capillaries*. In living tissue, grey matter actually has a grey-brown colour, which comes from capillary blood vessels and neuronal cell bodies. Grey matter includes regions of the brain involved in muscle control, sensory perception such as seeing and hearing, *memory*, *emotions*, and *speech*, being primarily associated with processing and cognition. This matter is distributed at the surface of the cerebral hemispheres (cerebral cortex) and of the cerebellum (cerebellar cortex); as well in the depth of the cerebrum (thalamus, hypothalamus, subthalamus, basal ganglia, nucleus accumbens, septal nuclei), cerebellar (deep cerebral nuclei - dentate nucleus, globose nucleus, emboliform nucleus, fastigial nucleus), brainstem (substantia nigra, red nucleus, olivary nuclei, cranial nerve nuclei), and also spinal grey matter (anterior horn, lateral horn, posterior horn).

Substantia nigra is a brain structure located in the mesencephalon (midbrain) that plays an important role in eye movement, motor planning, reward-seeking, learning, and addiction. Its black colour reflects the fact that its parts appear darker than neighbouring areas due to high levels of *melanin* in dopaminergic neurons. Substantia nigra consists of two parts: the *pars compacta* which serves mainly as an input to the basal ganglia circuit, supplying the striatum with dopamine; and the *pars reticulata* which serves mainly as an output, conveying signals from the basal ganglia to numerous other brain structures.

In more evolved vertebrates, the *brain stem* is most ancient part of brain, comprising the hindbrain plus the midbrain, minus the cerebellum. It is involved in alertness and in monitoring basic survival functions such as breathing, heartbeat, and blood pressure. The brain stem is also known as the 'reptilian brain' because it is considered the entire brain of reptiles. As a part of brain stem, the hindbrain is located at the rear of the skull, and includes: the medulla (controlling breathing, regulating reflexes, and monitoring the posture of the body), the cerebellum (two rounded structures, located besides the medulla, responsible for coordinating motor activity), and the pons (as a bridge towards the midbrain, which is responsible for monitoring sleep and arousal by coordinating with the autonomic nervous system). The other

27

part of the brain stem is the midbrain, which serves to relay information between the hindbrain and the forebrain, particularly information coming from the eyes and ears, and is composed of two systems: the reticular formation, involved with stereotypical patterns of behaviour such as walking, sleeping, and other reflexes; and a cluster of neurons having dopamine, serotonin, and norepinephrine receptors. The brains of vertebrates also include the optic tectum which allows actions toward points in space, responding to visual input, and directs eye movements, reaching movements and other object directed actions; and olfactory bulb which processes olfactory sensory signals and sends output to the olfactory part of the brain.

The *mammalian forebrain*, including the human one, is composed of the: *limbic system* (made up of the amygdala, responsible for processing emotions and involved in discrimination of objects, and the hippocampus, involved in memory storage and spatial memory); *thalamus* (located on top of the brain stem, and relaying information to different parts of the forebrain, as well as regulating the states of sleep and wakefulness); *hypothalamus* (located just below the thalamus, and involved in monitoring pleasurable activities such as eating, drinking, and sex, secreting hormones in response to different emotions, stress, and rewarding feelings, as well as being associated with autonomic elements of hunger and blood pressure such as cat hisses, raises tail, spits, extends claws, growls, piloerects and dilates pupils); *basal ganglia* (located between thalamus and the central cortex, and working with the cerebral cortex and the cerebellum to coordinate voluntary movements, particularly to form habitual behaviours, such as bicycle riding and typing); and *cerebral cortex* (the most recently developed part of the brain, which almost completely caps the rest of the brain parts). The cerebral cortex is divided into two hemispheres: the *left hemisphere* (in humans associated with verbal processing, such as speech and grammar, as well as mathematics) which processes information coming from the right side of the body; and the *right hemisphere* (involved with nonverbal processing, such as spatial perception, visual recognition, and emotion), which processes information coming from the left side of the body. These two hemispheres are connected with each other by a bundle of axons called 'corpus callosum'. Aside from being divided into two hemispheres, the cerebral cortex is also divided into occipital, temporal, parietal, and frontal *lobes*; and visual, auditory, motor, and sensory *cortices* as well as *association cortices*. Notably, the prefrontal cortex of the primates, including humans, and parts of their cortex are involved in vision,

28

carrying out functions that include planning, working memory, motivation, attention, and executive control.

In mammals, the optic tectum is called *superior colliculus*, a part of the midbrain, and the olfactory bulb is greatly reduced comparing with other vertebrates. More intelligent mammals such as dolphins, chimps and humans have highly convoluted brains.

The most obvious difference between the brains of mammals and other vertebrates is the size. On average, a mammal has a brain roughly twice as large as that of a bird of the same body size, and ten times as large as that of a reptile of the same body size. The *hindbrain* and *midbrain* of mammals are generally similar to those of other vertebrates, but noticeable differences appear in the *forebrain*, which is greatly enlarged and structured. The *cerebral cortex* is the part of the brain that most strongly distinguishes mammals. In non-mammalian vertebrates, the surface of the cerebrum is lined with a comparatively simple three-layered structure called the pallium; while in mammals, the pallium has evolved into a complex six-layered structure called the *neocortex*, or *isocortex*. Several areas at the edge of the neocortex, including the *hippocampus* and *amygdala*, are also much more extensively developed in mammals than in other vertebrates. The superior colliculus, which plays a major role in visual control of behaviour in most vertebrates, shrinks to a small size in mammals, and many of its functions are taken over by visual areas of the cerebral cortex. The *cerebellum* of mammals contains a large portion, namely the *neocerebellum*, dedicated to supporting functions of the cerebral cortex, which has no counterpart in other vertebrates.

In mammals, there is a nonlinearity of the brain-to-body relationship that has been expressed (R Holloway and D Post, 1982) by the *encephalization quotient*

$$EQ = [\text{brain weight}] \, / \, [(0.12) \cdot (\text{bodyweight})],$$

which indicates for some species values such as: *rat* 0.4; *horse* 0.9; *dog* 1.2; *elephant* 1.13 - 2.36; *bottlenose dolphin* 4.14; *Rhesus monkey* 2.1; *chimpanzee* 2.2 - 2.5; *human* 7.4 - 7.8.

As a distinct part of the peripheral nervous system in higher animals, the *autonomic nervous system* (ANS) plays a role in the body's control system, functioning largely below the level of consciousness, and governing visceral functions. This system affects the heart rate, digestion, respiratory rate, salivation, perspiration, pupillary dilation, micturition (urination), and sexual arousal. Within the brain, the ANS is located in the medulla oblongata of the lower brain stem. Its functions subdivide into other areas, and are also linked to ANS subsystems and nervous systems external to the brain. The hypothalamus, just above the brain stem, acts as an integrator for autonomic functions, receiving ANS regulatory input from the limbic system to do so. The ANS is divided into two complementary subsystems: the *parasympathetic nervous system*, a 'more slowly activated dampening', or 'rest and digest', or 'feed and breed' system; and the *sympathetic nervous system*, a 'quick response mobilizing', or 'fight or flight' system.

Communication development

Closely related to the life evolution, the communication within and between organisms evolved by cell signalling and intercellular communication, cell signalling in multicellular organisms, signalling molecules, circulation of fluids, communication in plants and animals, invertebrate communication, and communication in vertebrates.

An example of communication at micro- and nano-scales is evidenced by cells responding to changes in their environments through the bacterial DNA and RNA transcription from genes encoding enzymes which metabolize *arabinose* (an aldopentose, i.e. a monosaccharide containing five carbon atoms and including an aldehyde functional group) as the *mechanism of gene regulation*.

<u>Cell signalling and intercellular communication</u> are fundamental components of both plant and animal communication.

Cell signalling is part of a complex system of communication governing basic cellular activities and coordinating cell actions. The ability of cells to perceive and correctly respond to their microenvironment is also the basis of development, tissue repair, immunity, and normal tissue homeostasis. An important role in *intercellular communication* is played by *enzimes* (A Payen, 1833; L Pasteur, 1865-85; WF Kühne, 1876-77; E Buchner, 1897), which are indispensable for signal transduction and cell regulation, often via *kinases* (biochemical agents, such as a metal ion or a protein, which convert zymogens 'non-catalytic substances' to enzymes, acting as activators) and *phosphatases* (enzymes removing a phosphate group from its substrate), where phosphorylation of proteins by kinases is a remarkable mechanism in communicating signals within a cell (signal transduction) and regulating cellular activity, such as cell division. Enzymes are large biological molecules responsible for numerous metabolic processes that sustain life, acting as highly active catalysts and accelerating both the rate and specificity of metabolic reactions. Most enzymes are proteins, although some catalytic ribonucleic acid molecules have been also identified. In enzymatic reactions, the

30

molecules at the beginning of the process, called *substrates*, are converted into different molecules, called *products*. Unicellular and multicellular organism cell signalling is also significant in many mammals, where early embryo cells exchange signals with cells of the uterus.

During mating in the yeast *Saccharomyces cerevisiae*, some cells send a peptide signal (mating factor *pheromones*) into their environment, and the mating factor peptide may bind to a cell surface receptor on other yeast cells, inducing them to prepare for mating. Signalling within, between, and among cells is subdivided into intracrine, autocrine, juxtacrine, paracrine, and endocrine.

Cell signalling in multicellular organisms takes place between cells through either release into extracellular space, divided in *paracrine signalling* (over short distances) and *endocrine signalling* (over long distances); or by direct contact, known as *juxtacrine signalling*. *Autocrine signalling* is a special case of paracrine signalling where the secreting cells respond to the secreted signalling molecule (B Alberts *et al.*, 2002); and *synaptic signalling* is a special case of paracrine signalling (at chemical synapses) or juxtacrine signalling (at electrical synapses) between neurons and target cells. These kinds of signalling result in the activation of *second messengers*, leading to various physiological effects.

There are at least three important classes of *signalling molecules* widely recognized: *hormones* in the endocrine system; *neurotransmitters* in the nervous system, also including *neuropeptides* and *neuromodulators*; and *cytokines* in the immune system; with their growth factors; all of them being called *receptor ligands*.

Receptors for cell motility and differentiation can be exemplified by *Notch signalling* (a juxtacrine signalling) as a receptor, in which two adjacent cells must make physical contact in order to communicate; animals have a small set of genes coding for signalling proteins which interact specifically with Notch receptors and stimulate a response in cells that express Notch on their surface.

Circulation of fluids in higher animals is also significant to unveil the communication between cells and organs, by which the existence and vital functions of these organisms would be impossible. It includes the cardiovascular system, lymphatic system, and chyle, as well as the circulation of cerebrospinal, interstitial and extracellular fluids.

Cardiovascular system, also known as circulatory system, consists of the heart, described as a muscle functioning as a pump and effecting the

31

movement of the blood through body via the lungs by means of the arteries, the blood then returning through the veins to the heart (W Harvey, 1628). The blood itself comprises a connective tissue of liquid plasma, as fibrin and serum, and the corpuscles together with fibrin composing the clot, as a mass of soft or fluid matter.

Lymphatic system acts as a secondary circulatory system in vertebrates, and plays a crucial role in maintaining homeostasis as well as good health (T Bartholin, 1653-54), which consists of three main components, namely a complex capillary network that carries the lymph; collecting vessels that drain the lymph back into the bloodstream; and lymph glands 'nodes' that filter the lymph as it passes through. Unlike circulatory system, the lymphatic system is not closed and has no central pump, being a one-way system with fluid travelling from the interstitial space back to the blood, and, in most animals, a key physiological system for protection against invaders which have entered the body, such as bacteria, viruses and fungi.

Chyle is a fluid mixture of lymph and chylomicrons, consisting of lymph and emulsified fatty acids, which is formed in the small intestine and taken up by lymph vessels called *lacteals*; the chylomicrons being small fat globules composed of protein and lipids, which combine in the lining of the intestine, serving to transport fat from its port of entry in the intestine to the liver, and to adipose (fat) tissue.

Cerebrospinal fluid is produced in the choroid plexus of brain, and acts as a cushion or buffer for the cortex, providing basic mechanical and immunological protection of the brain inside the skull, and serving a vital function in autoregulation of cerebral blood flow (W Mestrezat, 1912; HW Cushing, 1914).

Interstitial fluid, or tissue fluid, is a solution that bathes and surrounds the cells of multicellular animals, representing the main component of the extracellular fluid which also includes plasma and transcellular fluid; by bathing the cells of the tissues, it provides a means of delivering materials to the cells, intercellular communication, and removal of metabolic wastes.

Extracellular fluid usually denotes all body fluid outside of cells, the remainder being called intracellular fluid; in some animals, including mammals, it can be divided into two major sub-compartments, namely interstitial fluid and blood plasma, also including transcellular fluid.

<u>*Communication in plants and invertebrates*</u> is also an essential attribute of life, taking place on its entire scale, including *cellular communication*, and *chemical transmissions* between very simple

organisms like bacteria and within the plant and fungal kingdoms. Inter-bacterial signalling can be exemplified by *Salmonella enteritidis* that uses acyl-homoserine lactone for *Quorum* sensing.

Other forms of communication are those existing between *fungi* to coordinate and organize their growth or development, e.g. formation of *Marcelia* (used in pharmaceutics as a powerful anti-viral extract) and fruiting bodies, and also with related species, non-fungal organisms in a large variety of symbiotic interactions, especially with bacteria, unicellular eukaryotes, plants and insects through biochemicals.

Plants of the same or related species communicate between them, or with animal organisms, especially in the roots by which interact with rhizome bacteria, fungi and insects in the soil, and also within and between plant cells. Parallel signalling in plants, as mediated communication processes, can be *intraorganismic* - within the plant body; *interorganismic* - between plants, and *transorganismic* - between them and non-plant organisms (G Witzany and B František, 2012). Such interorganismic and transorganismic communications can be exemplified by seeds, pollen, or ejections.

Communication plays a role in all information exchanged between living species, including plants and fungi. Transmission of signals involving a living sender and receiver takes place even in primitive creatures such as *corals* which are competent to communicate.

Capabilities of communication significantly increased after the appearance of *bilaterians*, which have a bilateral symmetry across and anterior-posterior (head-tail) axis. From their very beginning, the bilaterians differentiated into *protostomes*, with nerve chord running along the ventral side of the body; and *deuterostomes*, with nerve chord on the dorsal side of the body. Both protostomes and deuterostomes use similar genes to pattern their central nervous system, this fact being revealed by developmental studies to identify the ancestral animal that gave rise to the arthropod and mammalian lineages (EM De Robertis and Y Sasai, 1996), for which it is proposed the name *Urbilateria* (German *ur* or *urig* 'original'), as the first bilaterian organism with characteristics similar to the fossil *Kimberella*, dating to 555 million years ago. Thus the diverse bilaterian lineages use similar genes during the development of their central nervous systems or brains (A Ghysen, 2003; R Lichtneckert and H Reichert, 2005; A Denes *et al.*, 2007).

Invertebrate communication is widely spread on the Earth, taking into account that the invertebrates represent 96-98% from total animal species. Examples include *crabs* producing sounds that are transmitted via air and substrate, *marine invertebrates* producing sounds for defensive and courtship purposes, and *lobsters* generating stridulatory sounds via their plectrum and file when the antennae are moved. Male *fiddler crabs* wave their giant claw to attract female fiddler crabs.

Among the insects, some members of the social *ants* are on duty to alert other members of the colony during their sleep. Another example of insect communication is the *bees' dance*, when they find nectar, so directing other bees to the nectar location.

At similar levels of evolution, the mobility of animals is relatively higher than the mobility of plants, which implies more complex systems of communication between animals than between plants.

The development from simple through complex to sophisticated systems of communication between animals can be unveiled by several examples below.

Communication in vertebrates is distinguishable by their chemical and acoustic systems, including communication between a mother and her offspring, intercellular communication, and glial interactions between them and others.

Fishes communicate in many ways, each species having its own form of communication. Among them, the weakly *electric fishes* use a communicating channel, undetectable by other animals, namely the electric signalling, as an electric field generated by one fish and received by electroreceptors of another fish. Other fishes communicate with each other for attracting mates, scaring of predators, or orienting themselves.

On the scale of evolution, an important role of communication in *amphibians* and *reptiles* is played by *pheromones*, secreted or excreted chemical factors triggering social responses in members of the same species, which include alarm, food trial, and sex pheromones. *Garter snake* females attract males that chemically persuade a female to mate. Among the extinct dinosaurs, *Parasaurolophus* was able to emit infrasounds for communicating between them or for keeping the predators away. Another dinosaur, the night predator *Ankylosaurus* had a very developed sense of smell.

Birds vocally communicate through calls either with other individuals of the same species or across species, and are able to produce a variety of sounds to communicate with flock members, mates, neighbours and family members. *Swans* entwine their long necks both to fight and to court.

Marsupials also communicate between them, for example *kangaroos* thump their hind legs to warn others of danger.

Communication between *mammals* takes place by modes such as:

visual (gestures, facial expression, gaze by coordinating social animals, active and passive visual displays, and bioluminescent communication), *auditory* (through vocalizations), *olfactory* (by glands that generate distinctive and long-lasting smells), *electrocommunication* (especially in aquatic animals, and land mammals, notably *platypus* and *echidnas*), *seismic* (vibrational communication, conveying information through vibrations of the substrate), and *autocommunication* (by the same animal as sender and receptor, including echolocation in *bats* and dolphins, and active electrolocation in weakly electric fish). Among the mammals there are more than 30 species of *chatty dolphins*, which have a 'secret language' (primarily used for communication between mothers and their babies); and the *bottlenose dolphins* which develop distinctive high-pitched whistles, serving for individual identification, and for keeping contact between them or for preying. The social mammal *meerkats* emit, in case of danger, various calls with specific meaning which indicate the type of predator and the urgency of the situation; as well as alarm, panic, recruitment and moving calls. In meerkat system of communication, there were identified calls with six different meanings: for aerial predator - low, medium, and high urgency; and for terrestrial predator - low, medium, and high urgency. Other examples of mammal communication are: *dogs* stretch their front legs out in front of them and lower their bodies when they want to play; *prairie dogs* bare their teeth and press their mouths together to discover if they are friends or foes; *white-tailed deer* indicate alarm by flicking up their tails; *elephants* show affection by entwining their trunks; *giraffes* press their necks together when they are attracted to each other; *horses* rub noses as a sign of affection; and *whales* breach (leap out of the water) repeatedly to send messages to other whales.

In natural conditions, the <u>*animal communication*</u> consists of: *chemical signals* (used by very simple organisms including protozoa), *smell* (such as pheromones attract, skunk secretions repel), *touch* (direct contact of bodies), *movement* (act or manner of changing position), *posture* (e.g. geese, dogs), *facial gestures* (such as dogs snarling, or their 'play face' and tail signals), *visual signals* (e.g. bird feathers), and *sound* (such as invertebrate and vertebrate calls). *Signals* used for communication can be instinctive (genetically programmed), or learned from others; and evolved in order to *attract* (such as in mating), *repel* (competitors or enemies), *aggress or submit*, *advertise* species, *warn* of predators, *inform* about danger or availability of food.

Animal communication is usually differentiated by its _modes_: *(i) visual*, e.g. gestures, facial expression, gaze, displays, bioluminescent communication (some bacteria, protists, comb jellies, crustaceans, cephalopods, click beetles, glow worms, anglerfish); *(ii) auditory*, in cases of hammer-headed bats, red deer, humpback whales, elephant seals, prairie dogs, gibbons, and vervet monkeys; *(iii) olfactory*, such as Mongolian gerbils, some fish, golden hamsters, cats; *(iv) electric*, e.g. platypus and echidna electroreception; *(v) seismic*, such as vibrational communication in crustaceans, arachnids, insects, amphibians, reptiles, birds, mammals, birds; and *(vi) autocommunication*, including echolocation in bats and dolphins, electrolocation in weakly electric fish.

Energy of signalling and communication E_{sc} used by an individual organism, both inside and outside of its body, ranges widely from the smallest *yeast* or *virus* to the biggest *Populos tremuloides* or *blue whale*. The difficulty to assess the energy E_{sc} can be passed over by correlating it with the _energy of growth_ E_{gr} that varies roughly from $5 \cdot 10^{-23}$ joules for the smallest organisms to $5 \cdot 10^{-15}$ joules for the biggest organisms (M Albu, 2013-14). Assuming that between 10% and 50% from the growth energy of an organism is used for signalling-communication, the order of magnitude of its signalling-communication energy is approximated as shown below.

Organism	Growth energy E_{gr} [J]	Percentage of growth energy used for signalling-communication [%]	Signalling-communication energy E_{sc} [J]
Smallest	$5 \cdot 10^{-23}$	10	$5 \cdot 10^{-24}$
		20	$1 \cdot 10^{-23}$
		50	$2.5 \cdot 10^{-23}$
Biggest	$5 \cdot 10^{-15}$	10	$5 \cdot 10^{-16}$
		20	$1 \cdot 10^{-15}$
		50	$2.5 \cdot 10^{-15}$

1.3. Timescale processing

During the early times of Earth's history, the evolution of life was marked by living forms and organisms, such as *non-cellular living forms*, able to synthesize proteins independent of a host cell and a cellular membrane, which emerged from 4200-4000 million years ago. They were followed by *proto-organisms with ribosomal RNA genes*, as unicellular micro-organisms with DNA lying freely in cytoplasm, reproducing by binary fission, fragmentation, budding, or spores, and including primitive forms of *archaea*; and *protokaryotes*, as unicellular organisms whose cells lack a membrane-bound nucleus 'karyon', as well as mitochondria, or any other membrane-bound organelles, their cells being located together in an area enclosed by a wall made up of peptidoglycin, rather than separated in different cellular compartments, including early *bacteria*. Notably, such organisms had an amazing ability to live in extreme environments, because each of them generates an endospore by copying its chromosome, producing a very tough structure around it that can resist boiling and remain dormant for hundreds years until their environment improves, when they can rapidly reproduce and subsist because of their high adaptive immunity. However, these first forms of life were able to send and receive simplest *signals* not only within themselves but also to interact, and thus to make possible the *first kind of communication* on the Earth.

Although oxygen was undoubtedly released by photosynthesis well back in Archaean times, it could not build up to any significant degree until chemical sinks, unoxidized sulphur and iron, had been filled. Thus, until around 2300 million years ago, oxygen was probably only 1-2% of its current level. Banded iron formations indicate an increase in atmospheric oxygen after about 2000 million years ago. The oxygen build-up was mainly due to two factors: a filling of the chemical sinks; and an increase in carbon burial which sequestered organic compounds that would have otherwise been oxidized by the atmosphere.

Beginning with the simply organized unicellular filamentous blue-green algae, plants became able to process their nutrients, in presence of sunlight, by withdrawing carbon dioxide from and releasing oxygen to the atmosphere, thus changing the atmospheric composition by photosynthesis and sustaining the evolution of entire life on our planet. The Great Oxidation Event took place between 2500 and 1800 million years ago, when the Earth's *third atmosphere* evolved by transferring carbon dioxide to and from continental carbon stores during the rearrangement of continents by plate tectonics, and by appearance of free oxygen from about 1800 million years ago indicated by the end of banded iron formations consuming oxygen. The oxygen in this third atmosphere exceeded the capacity of consumption by reducible materials, so that the amount of oxygen increased even more, perhaps up to 25-30%, significantly higher than today's 21%; the changes in composition of the atmosphere were governed by two main processes, namely the use of carbon dioxide from and release of oxygen to the atmosphere by plants' photosynthesis, and the oxidation of pyrite and volcanic release of sulphur into the atmosphere which tended to reduce the content of atmospheric oxygen.

In early times, *stromatolites*, stubby pillars built by colonies of microorganisms, played a major role in environment changes and in life evolution, but their abundance and diversity declined after 1300-1200 million years ago, because of grazing and burrowing animals. These animals evolved from acritarchs and underwent a boom around 1000 million years ago, increasing in abundance, diversity, complexity of shape, and especially size and number of spines which made them *predators* and imposed upon their preys a development of defensive means. Such predatory metazoans evolving between 1000 and 700 million years ago were the predecessors of Cambrian animals which by their complexity determined the further course of entire life on the Earth (S Bengtson, 2002; S Poter, 2011).

In Phanerozoic, plants by their photosynthesis continued to use carbon dioxide from and to release oxygen into atmosphere, so that the evolution of life became faster and diversified. Subsequently, this carbon dioxide was stored in Cretaceous formations known around the world, including limestone that makes up about 10% of the total volume of all sedimentary rocks.

The global sum of all ecosystems, as the zone of life on Earth, represents a closed and self-regulating system, called *biosphere*, which extends roughly from 11,000 metres below the ocean's surface to 11,000 metres in the atmosphere. In modern biology, there are five *unifying principles*:

1st - Cells are the basic units of life;

2nd - New species and inherited traits result from evolution;

3rd - Genes are the basic units of heredity;

4th - An organism regulates its internal environment to maintain stable and constant conditions; and

5th - Living organisms consume and transform energy.

In the current biological classification, there is in use a hierarchy of *eight major taxonomic ranks*: domain, kingdom, phylum, class, order, family, genus, and species. Today, the number of identified species is over 6.5 million on land and 2.2 million in ocean.

The development of communication within and between organisms was intrinsically related to the life evolution, and therefore their timeline and sequences are closely corresponding to each other. Starting from around 4200 million years ago and continuing until today, the signalling and communication within and between organisms developed and differentiated, taking place *by senses* (touch, taste, smell, hearing, sight) *through material or empty spaces* (air, water, other fluids, ground, void), *involving various forms of energy* (e.g. thermal, chemical, mechanical, electrical, radiant) used for procreation, multiplication, reproduction, growth, defence, attack, etc., and ultimately for surviving and perpetuating species.

Identified time intervals in evolution of life
Based on anatomical and physiological characteristics, geological and paleontological studies, fossil records and genetic lineages, dating and molecular clock techniques, climatic and ecologic changes, biologists are reconstructing and updating a timeline for the evolution of life on the Earth, which currently consists of intervals of time when main groups of plants and animals emerged, in million years ago, as follows:

4200 - 3800: *Protokaryotes, archeons, early bacterial forms*;

3800 - 3600: *Microbial mat forms, bacteria*;

3600 - 3400: *Prokaryotes, archaea*;

3400 - 2000: *Cyanobacteria, acritarchs*;

2000 - 1800: *Eukaryotic cells*;

1800 - 1400: *Stromatolites, eukaryotes*;

1400 - 1000: *Multicellular organisms, red algae, dinoflagellates*;

1000 - 700: *Metazoans, vaucherian algae*;

700 - 500: *Early fungi, bilaterians, chordates, arthropods, molluscs, brachiopods, foraminifers, radiolarians, graptolites, cephalopods, chitons*;

500 - 450: *Sponges, corals, anemones, jawless fish, land plants*;

450 - 400: *Conodonts, echinoids, fishes, arachnids, land scorpions*;

400 - 350: *Insects, lichens, hexapods, ammonoids, sharks, tetrapods, amphibians, crabs, ferns*;

350 - 300: *Ratfish, hagfish, amniote vertebrates, synapsids*;

300 - 200: *Reptiles, beetles, therapsids, early ichthyosaurus, early* *dinosaurs, bivalves, teleost fishes, first mammals, gymnosperms,* fossil *polen grains* from earliest flowering plants, *flies, turtles;*

200 - 150: *Mammals, pterosaurs, pliosaurs, first lepidopteran insects, stegosaurs, salamanders, newts, plesiosaurs, cladotherian mammals, Archaeopteryx, triconodontid and symmetrodont mammals;*

150 - 100: *Birds, angiosperms, freshwater turtles, silicoflagellates, diatoms, first monotreme mammals;*

100 - 50: *Early bees, snakes, ticks, first ants, flightless birds, primates, carnivorous mammals, owls, modern birds, whales, proboscideans, rodents, first bats;*

50 - 20: *Tapirs, rhinoceroses, camels, butterflies, moths, grasses, dogs, eagles, hawks, marsupials, first balanids, eucalypts, pigs, cats, deer;*

20 - 10: *Great apes* (Hominidae), *giraffes, hyenas, bears, bovids, kangaroos;*

10 - 5: *Horses, hominines;*

5 - 0.2: *Hippopotami, big vultures, mammoths, elephants, zebras, lions, gazelles, Australopithecines, Paranthropus, genus Homo, Homo antecessor, Homo heidelbergensis, Homo neanderthalensis, anatomically modern humans.*

Time function for natural processes

The intervals of time above are succeeding each other in a seemingly regular manner, rather than randomly, that indicates an overall tendency of decreasing their durations, i.e. shortening in course of life evolution on the Earth. This tendency is notable in course of any natural process, for instance the Earth's history itself is marked by shortening eras such as: Palaeoproterozoic 2500 - 1600, Mesoproterozoic 1600 - 1000, Neoproterozoic 1000 - 541, Palaeozoic 541 - 252, Mesozoic 252 - 66 million years ago.

Although its great scale, diversity and complexity, the evolution of life is a natural process and therefore can be analyzed on the whole taking into account that: *(i)* its energy derives from surrounding sources (e.g. thermal, chemical, mechanical, electromagnetic, organic) and differentiates into a *t-energy* E_t changing with increasing time t, and a complementary *τ-energy* E_τ changing with decreasing period τ, so that the total energy $E = E_t + E_\tau$ is constant on the entire course of evolution from an initial time $t_o = 0$ to a final time $t_\bullet$; and *(ii)* its efficiency, measured by the ratio of *t-power* P_t to *τ-power* P_τ, increases directly proportional to the running time t, but must always be less than 100% which would imply perpetual process.

40

Therefore, the *total energy E* [joules] of evolution is a function of both time t and period τ, i.e. $E = E(t, \tau)$, with the total differential $dE = (\partial E/\partial t) \cdot dt + (\partial E/\partial \tau) \cdot d\tau = P_t \cdot dt + P_\tau \cdot d\tau$, where the terms $(\partial E/\partial t) \cdot dt$ and $(\partial E/\partial \tau) \cdot d\tau$ express the infinitesimal changes in *t-energy* and in *τ-energy* respectively; whilst their corresponding partial derivatives $\partial E/\partial t = P_t$ and $\partial E/\partial \tau = P_\tau$ express the instantaneous *t-power* and *τ-power* [watts]. As the total energy E is constant on the entire course of evolution, it follows that its total differential is always nil, so that $dE = P_t \cdot dt + P_\tau \cdot d\tau = 0$, or $P_t \cdot dt = -P_\tau \cdot d\tau$, and then the quotient $Q = P_t/P_\tau$ is given by the equality

$$Q = -d\tau/dt.$$

Representing the efficiency of evolution, the quotient Q increases directly proportional to the time t, i.e. $Q = $ (constant)$\cdot t$, or $Q/t = $ constant, and therefore $d(Q/t) = 0$ or $(t \cdot dQ - Q \cdot dt)/t^2 = 0$, whence, for t finite, results the equation $t \cdot dQ - Q \cdot dt = 0$ or $dQ/Q = dt/t$. As $Q \leq 1$ and $t \leq t_\bullet$, the equation $dQ/Q = dt/t$ leads, through integration $\int_{Q}^{1} dQ/Q = \int_{t}^{t_\bullet} dt/t$, to $ln(1) - ln(Q) = ln(t_\bullet) - ln(t)$, or $ln(Q) = ln(t/t_\bullet)$, and then to the equality

$$Q = t/t_\bullet.$$

Two above equalities result in the equation $-d\tau/dt = t/t_\bullet$, or $-d\tau = (1/t_\bullet) \cdot t \cdot dt$ that, integrated as $-\int_{\tau}^{0} d\tau = (1/t_\bullet) \cdot \int_{t}^{t_\bullet} t \cdot dt$, discloses an expression of the period

$$\tau = (t_\bullet^2 - t^2)/(2t_\bullet),$$

where $0 \leq t \leq t_\bullet$; $t_\bullet/2 \geq \tau \geq 0$.
On the other hand, the curly *τ-energy* is cycling at a rate of decreasing period τ, or equivalently of increasing frequency v, that corresponds to an instantaneous angular speed ω defined by the time rate of an ever increasing angle θ, such as $\omega = d\theta/dt = 2\pi/\tau = 2\pi \cdot v$, whereby the equality $d\theta = (2\pi/\tau) \cdot dt$, and then another expression of the period

$$\tau = 2\pi \cdot dt/d\theta.$$

The two complementary expressions of period established above lead to the equality $(t_\bullet^2 - t^2)/(2t_\bullet) = 2\pi \cdot dt/d\theta$, and thus to the differential equation $d\theta = 4\pi \cdot t_\bullet \cdot dt/(t_\bullet^2 - t^2)$, or $d\theta = 4\pi \cdot d(t/t_\bullet)/[1 - (t/t_\bullet)^2]$, that, integrated as $\int_{0}^{\theta} d\theta = 4\pi \cdot \int_{0}^{t/t_\bullet} d(t/t_\bullet)/[1 - (t/t_\bullet)^2]$, results in the function $\theta(t) = 4\pi \cdot tanh^{-1}(t/t_\bullet)$, or

$$\theta/4\pi = tanh^{-1}(t/t_\bullet),$$

where the angle θ [radians] is given by the product of a double-cycle $4\pi = 2 \cdot (2\pi)$ and the inverse hyperbolic tangent $tanh^{-1}$ of the ratio of running time t to final time $t_\bullet$. The inverse of $\theta/4\pi = tanh^{-1}(t/t_\bullet)$ is just the useful *time function*

$$t = t_\bullet \cdot tanh(\theta/4\pi),$$

41

in which $tanh(\theta/4\pi)$ is the hyperbolic tangent of the increasing angle θ divided by a double-cycle 4π, and t is the time running from $t_\circ = 0$ to $t_\bullet$ (M Albu, 2013-14).

Taking into account that the process of evolution is lasting billions of years and its subdivisions could last from hundreds million to tens of years, it is convenient that from the multitude of values taken by the time function to be selected only those corresponding to values of argument included in one of the following sets of order:

Zero-order $\theta/4\pi = N = 0, 1, 2, 3$, etc. as integers;

First-order $N/10 = \alpha$ as tenth-integers;

Second-order $N/10^2 = \beta$ as hundredth-integers;

Third-order $N/10^3 = \gamma$ as thousandth-integers;

Fourth-order $N/10^4 = \delta$ as ten thousandth-integers;

Fifth-order $N/10^5 = \varepsilon$ as hundred thousandth-integers;

Sixth-order $N/10^6 = \zeta$ as millionth-integers;

Seventh-order $N/10^7 = \eta$ ten millionth-integers.

These sets will be used to calculate the discrete times delimiting every kind of sequences during the evolution.

Biological timescale processing

At the full scale, the evolution of life can be processed taking into account that: *(i)* the first non-cellular forms of life originated around 4200 million year ago, as its initial time $t_0 = 0$; and *(ii)* the representative groups of organisms successively emerged at the approximate times, in million years ago, such as

\ 4200 \ *Protokaryotes*

\ 3820...3700 ≈ 3760 \ *Bacteria*

\ 3380...3280 ≈ 3330 \ *Prokaryotes*

\ 2970...2870 ≈ 2920 \ *Cyanobacteria*

\ 2570...2490 ≈ 2530 \ *Acritarchs*

\ 2200...2135 ≈ 2167 \ *Unicellular eukaryotes*

\ 1865...1805 ≈ 1835 \ *Stromatolites*

\ 1570...1510 ≈ 1540 \ *Algae*

\ 1300...1255 ≈ 1278 \ *Dinoflagellates*

\ 1070...1030 ≈ 1050 \ *Multicellular organisms*

\ 870...830 ≈ 850 \ *Metazoans* and *vaucherians*

\ 690...665 ≈ 678 \ *Cephalopods*

\ 545...520 ≈ 532 \ *Fish*

\ 420...395 ≈ 408 \ *Amphibians*

\ 315...295 ≈ 305 \ *Reptiles*

\ 225...210 ≈ 217 \ *Mammals*

\ 150...140 ≈ 145 \ *Birds*

\ 86...82 ≈ 84 \ *Primates* and *marsupials*

\ 35...33 ≈ 34 \ *Great apes*.

Using the first-order $N/10 = \alpha$ sequences, and converting the times above given in million years ago to times $t_\alpha = tanh(\alpha)$ from the origin of life, their succession is rewritten as:

$$t_{0.1} = t_\bullet \cdot tanh(0.1) \approx 4200 - 3760 \approx 440;$$

$$t_{0.2} = t_\bullet \cdot tanh(0.2) \approx 4200 - 3330 \approx 870;$$

$$t_{0.3} = t_\bullet \cdot tanh(0.3) \approx 4200 - 2920 \approx 1280;$$

$$t_{0.4} = t_\bullet \cdot tanh(0.4) \approx 4200 - 2530 \approx 1670;$$

$$t_{0.5} = t_\bullet \cdot tanh(0.5) \approx 4200 - 2167 \approx 2033;$$

$$t_{0.6} = t_\bullet \cdot tanh(0.6) \approx 4200 - 1835 \approx 2365;$$

$$t_{0.7} = t_\bullet \cdot tanh(0.7) \approx 4200 - 1540 \approx 2660;$$

$$t_{0.8} = t_\bullet \cdot tanh(0.8) \approx 4200 - 1278 \approx 2922;$$

$$t_{0.9} = t_\bullet \cdot tanh(0.9) \approx 4200 - 1050 \approx 3150;$$

$$t_1 = t_\bullet \cdot tanh(1) \approx 4200 - 850 \approx 3350;$$

$$t_{1.1} = t_\bullet \cdot tanh(1.1) \approx 4200 - 678 \approx 3522;$$

$$t_{1.2} = t_\bullet \cdot tanh(1.2) \approx 4200 - 532 \approx 3668;$$

$$t_{1.3} = t_\bullet \cdot tanh(1.3) \approx 4200 - 408 \approx 3792;$$

$$t_{1.4} = t_\bullet \cdot tanh(1.4) \approx 4200 - 305 \approx 3895;$$

$$t_{1.5} = t_\bullet \cdot tanh(1.5) \approx 4200 - 217 \approx 3983;$$

$$t_{1.6} = t_\bullet \cdot tanh(1.6) \approx 4200 - 145 \approx 4055;$$

$$t_{1.7} = t_\bullet \cdot tanh(1.7) \approx 4200 - 84 \approx 4116;$$

$$t_{1.8} = t_\bullet \cdot tanh(1.8) \approx 4200 - 34 \approx 4166.$$

As the ancestors of earlier organisms are dated less accurately than the ancestors of later ones, the later times only are chosen to estimate the final time $t_\bullet = t_\alpha/tanh(\alpha)$, as for example

$t_\bullet = t_{1.5}/tanh(1.5) = t_{1.6}/tanh(1.6) = t_{1.7}/tanh(1.7) = t_{1.8}/tanh(1.8)$,

whence

$t_\bullet \approx (3983)/(0.9051483) \approx (4055)/(0.9216686) \approx (4116)/(0.9354091) \approx$

$\approx (4166)/(0.9468060) \approx 4400$ million years from the origin of life.

According to this value of the final time, all times $t_\alpha = t_\bullet \cdot tanh(\alpha)$ delimiting the sequences of life evolution can be calculated and displayed as in the table below.

$N/10 = \alpha$	Time (million years)		Sequences in evolution of organisms
	from origin $t_\alpha = t_\bullet \cdot tanh(\alpha)$	from present $t_\alpha - 4200$	
	4400	+200	
...	...	...	
1.9	4207.4	+7.4	_
1.8	4165.9	-34.1	_Great apes
1.7	4115.8	-84.2	_Primates and marsupials
1.6	4055.3	-144.7	_Birds
1.5	3982.7	-217.3	_Mammals
1.4	3895.5	-304.5	_Reptiles
1.3	3791.6	-408.4	_Amphibians
1.2	3668.1	-531.9	_Fish
1.1	3522.2	-677.8	_
1	3351.0	-849.0	_Cephalopods
0.9	3151.7	-1048.3	_Metazoans, vaucherians
0.8	2921.8	-1278.2	_Multicellular organisms
0.7	2659.2	-1540.8	_Dinoflagellates
0.6	2363.0	-1837.0	_Algae
0.5	2033.3	-2166.7	_Stromatolites
0.4	1671.8	-2528.2	_Unicellular eukaryotes
0.3	1281.8	-2918.2	_Acritarchs
0.2	868.5	-3331.5	_Cyanobacteria
0.1	438.5	-3761.5	_Prokaryotes
0	0	-4200	_Bacteria
			_Protokaryotes

References

1543: <u>Andreas Vesalius</u>, Belgian anatomist, produced the great work *De Humani Corporis Fabrica* 'On the Structure of the Human Body', with excellent descriptions and drawings of bones, *nervous system*, and *thalamus*.

1628: <u>William Harvey</u>, English physician, first described completely and in detail the *systemic circulation* and properties of *blood* being pumped to the brain and body by the heart; the results of his work were presented in the book *De Motu Cordis*, otherwise known as 'On the Motion of the Heart and Blood'.

1653-54: <u>Thomas Bartholin</u>, Danish physician, mathematician and theologian, became well-known for discovery of the *lymphatic system* in animals and humans, as described in his works *Vasa lymphatica nuper Hafniae in animalibus inventa et hepatis exsequiae*, and *Vasa lymphatica in homine nuper inventa*, Copenhagen.

1664-90: <u>John Locke</u>, English empiricist philosopher, working for his books *Questions Concerning the Law of Nature*, and *An Essay Concerning Human understanding*, he introduced into English the term *semiotics* as a synonym for 'doctrine of signs'.

1778-79: <u>William Cullen</u>, Scottish chemist and physician, produced *First Lines of the Practice of Physic*, emphasizing the importance of the nervous system in causation of disease, and coining the word *neurosis* to describe a group of nervous diseases.

1833: <u>Anselme Payen</u>, French chemist, working to develop the processes for sugar refining, he discovered the first enzyme, which was called *diastase*.

1856-67: <u>Gregor Johann Mendel</u>, German-Czech Austrian monk and botanist, observed that organisms inherit traits via *discrete units of inheritance*, later called genes; as well as formulated *Mendel's first law* 'law of segregation', and *Mendel's second law* 'law of independent assortment', representing *Mendelian inheritance*, as basic concepts of modern genetics.

1862-71: <u>Charles Robert Darwin</u>, English naturalist, left invaluable knowledge regarding variation and interbreeding in organisms, and formulated the theory of *pangenesis*, concerning acquired and inherited characteristics.

1865-85: <u>Louis Pasteur</u>, French chemist, discovered that *fermentation* was catalyzed by a vital force contained within the yeast cells, called 'ferments', and it is essentially due to organisms and not spontaneously generated, thus founding the 'germ theory of disease', and showing that *disease is communicable through the spread of micro-organisms*.

1867-1902: <u>August Friedrich Leopold Weismann</u>, German biologist, investigated the development of the two-winged flies *Diptera*, describing its neurohumoral organ known as the *Weismann ring*; stated the *germ-plasm theory*, deducing that information required for development and final form of an organism must be contained within germ cells, eggs and sperm, and be transmitted unchanged from generation to generation; as well as published

Essay upon Heredity and Kindred Biological Problems, and *Vorträge über Descendenztheorie* 'Lectures on Evolutionary Theory'.

1873-1918: <u>Camillo Golgi</u>, Italian cytologist, discovered how to stain *nerve tissue using silver nitrate*, and *Golgi bodies* in animal cells which are then readily visible under microscope, thus opening a new field of research into the *central nervous system, sense organs, muscles*, and *glands*.

1874: <u>Addison Emery Verrill</u> and <u>Sidney Irving Smith</u>, US zoologists, wrote the volume *Report upon the Invertebrate Animals of Vineyard Sound*, as an important reference manual of Atlantic Coast marine invertebrate fauna.

1874-1936: <u>Ivan Petrovich Pavlov</u>, Russian physiologist, worked on three main areas of physiology, namely the *circulatory system, digestive system*, and *higher nervous activity* including the brain; made famous research showing that if a bell is sounded whenever food is presented to a dog, it will eventually begin to salivate when bell is sounded without food being presented, this process being termed *conditioned* or *acquired reflex*, and leading to theories of animal and human behaviour.

1876-77: <u>Wilhelm Friedrich Kühne</u>, German physiologist, researching the physiology of muscle and nerve, he discovered the protein-digesting enzyme *trypsin*, and first used the term *enzime* (Greek ενζυμον 'in leaven').

1892-1914: <u>Santiago Ramón y Cajal</u>, Spanish physician and histologist, worked on the microstructure of nervous system, revealing *how nerve impulses are transmitted to brain*, and wrote books including the two-volume *Estudios sobre la degeneración y regeneración del sistema nervioso* 'The Degeneration and Regeneration of the Nervous System'.

1897: <u>Eduard Buchner</u>, German chemist and zymologist, made experiments to produce a cell-free extract of yeast cells, showing that this 'press juice' *could ferment sugar*, i.e. sugar was fermented *even in absence of living yeast cell* in the mixture.

1900-20: <u>Charles Henry Turner</u>, US behavioural scientist, pioneered the field of *insect behaviour*, and became well-known for his work showing that *social insects* can modify their behaviour as the result of experience.

1905-06: <u>William Bateson</u>, English geneticist, studied the process of *linkage*, by which some genes are inherited together; and coined the word *genetics* (Greek γενετικός 'genitive or generative' in turn from γένεσις 'origin').

1906-19: <u>Charles Scott Sherrington</u>, English physiologist, studied the structure and function of nervous system, and analyzed reflexes, the results being summarized in *The Integrative Action of the Nervous System*; coined the word *synapse* (Greek σύν 'together' and άπτειν 'to seize, grasp') to describe the junction between nerve cells; and mapped the motor areas of the cerebral cortex of mammals, producing the influential textbook on experimental physiology entitled *Mammalian Physiology*.

1909: <u>Wilhelm Johannsen,</u> Danish botanist, coined the word *gene* (Greek γενος 'birth, race') to describe the fundamental physical and functional *unit of heredity*.

46

1911-13: <u>Alfred Henry Sturtevant</u>, US geneticist, produced the first chromosome map and the mathematical background for genetic mapping experiments on the fruit-fly *Drosophila*, as well as used the phenomenon of genetic linkage to show that *genes are arranged linearly on chromosomes*.

1912: <u>William Mestrezat</u>, French oenologist, physician and neurochemist, gave the first accurate description of the *chemical composition of the cerebrospinal fluid*.

1914: <u>Harvey William Cushing</u>, US neurosurgeon, published conclusive evidence that the *cerebrospinal fluid* is secreted by the *choroid plexus*.

1920: <u>Hans Winker</u>, German botanist, coined the term *genome* (German *Gen* 'gene' and *(Chromos)om* 'chromosome'), which was later defined as 'all DNA in a cell' because this includes not only genes but also DNA that is not part of a gene, or non-coding DNA.

1920-35: <u>Jakob Johann von Uexkül</u>, Estonian-German biologist, worked in the fields of muscular physiology, animal behaviour, and cybernetics of life, pioneering the *semiotic biology*; and wrote *Streifzüge durch die Umwelten von Tieren und Menschen* 'A stroll through the environments of animals and humans', where the term 'umwelt' was defined as the perceptual world in which an organism exists and acts as a subject.

1940-46: <u>Max Delbrück</u>, German biophysicist, worked on the *genetics of phage virus*, a simple organism with a protein coat surrounding a coil of DNA, and discovered that *viruses can exchange genetic material to create new types of virus*.

1942: <u>George Wells Beadle</u>, US biochemical geneticist, and <u>Edward Lawrie Tatum</u>, US biochemist, demonstrated the role of genes in biochemical processes by growing bread mould spores on a variety of nutritional media, and suggesting that each spore had one or more blocks in metabolic pathway for particular nutrients, which led to the *one gene, one enzyme* hypothesis, as a single gene codes for synthesis of one protein.

1944: <u>Oswald Theodore Avery</u> and <u>Maclyn McCarty</u>, US bacteriologists and geneticists, showed that a non-virulent rough-coated strain of bacterium could be transformed into a virulent smooth strain in presence of some dead bacteria, leading to *identification of molecule responsible for transformation* as (later called) DNA.

1945-61: <u>Harold Clayton Urey</u>, US chemist, apart from his previous discovery of deuterium, he worked with SL Miller for synthesizing organic molecules, and also on geochemistry and cosmic chemistry, writing *Cosmochemical Problems*.

1948-49: <u>Claude Elwood Shannon</u>, US mathematician, electronic engineer and cryptographer, published his article 'A mathematical theory of communication', becoming known as 'the father of information theory'; which was then developed, together with <u>Warren Weaver</u>, US scientist and mathematician, into a book entitled *The Mathematical Theory of Communication*, Urbana: University of Illinois Press.

1948-67: <u>Severo Ochoa</u>, US geneticist, studied the *energetics of carbon dioxide fixation in photosynthesis*; achieved the first *synthesis of artificial RNA*, solved the *amino acid genetic code*, and identified a number of *base triplets*; as well as studied the *direction of protein synthesis along DNA*, and the *first amino acid in a peptide sequence.*

1950-65: <u>François Jacob</u> and <u>Jacques Lucien Monod</u>, French biochemists, with <u>André Michel Lwoff</u>, French microbiologist, showed that *genes are controlled by a system of other genes which regulate certain enzymes*, and developed the *theory of operon system*, whereby a *regulator gene controls other genes* by binding to a specific section of the DNA strand.

1950-75: <u>Edward Lawrie Tatum</u>, US biochemist, and <u>Joshua Lederberg</u>, US biologist and geneticist, described the *transduction in bacteria*, whereby a bacterial virus transfers part of its DNA into a host bacterium, leading to the development of techniques for *manipulation of genes*; and evidenced the *sexual process of conjunction* in bacteria reproduction.

1952: <u>Alfred Day Hershey</u> and <u>Martha Chase</u>, US biologists, made the *Hershey-Chase experiment* showing that DNA, rather than protein, represents the genetic material of viruses infecting bacteria; and confirmed that *DNA of other organisms fulfils same role.*

1953: <u>Francis Harry Compton Crick</u>, English molecular biologist, and <u>James Dewey Watson</u>, US biologist, worked on the *structure of DNA* (deoxyribonucleic acid), finding that the biological molecule contained in cells carries genetic information; elaborated the famous model of a *double-helical molecule*, consisting of two strands of nucleotide bases wound around a common axis in opposite directions; and suggested a simple method for duplication, i.e. if strands are separated, new partner strands are reconstructed for each based on sequence of old strand.

1953-60: <u>Stanley Lloyd Miller</u>, US chemist, concerning the possible origins of life on Earth, he passed electric discharges (simulating thunderstorms) through mixtures containing reducing gases (hydrogen, methane, ammonia) and water, which resulted in *complex organic molecules.*

1954-58: <u>William Howard Stein</u> and <u>Stanford Moore</u>, US biochemists, developed a column chromatographic method for identification and quantification of *amino acid mixtures* in proteins and physiological tissues, analyzed the *base sequence of RNA*, and studied *novel protease from streptococcus*, showing that its molecular structure differed from that of plant protease papain, thus giving a first example of the phenomenon called *convergent evolution.*

1956: <u>George Emil Palade</u>, Romanian-born US cell biologist, developed a method of separating cell components, known as *cell fractionation*, and identified these components as *mitochondria, endoplasmatic reticulum, Golgi apparatus*, and *ribosomes*, showing that *protein synthesis occurs on strands of RNA in ribosomes.*

48

1959-2002: <u>Arthur Kornberg</u>, US biochemist, carried out biochemical researches especially on enzyme chemistry, DNA replication, and nucleic acids controlling heredity in animals, plants, bacteria and viruses; discovered *DNA polymerase*, an enzyme synthesizing new DNA; made a *synthesis of viral DNA*; and published books including *For the Love of Enzymes: The Odyssey of a Biochemist*, Harvard University Press, and *The Golden Helix: Inside Biotech Ventures*, University Science Books.

1966: <u>Marshall Warren Nirenberg</u> and <u>Robert William Holley</u>, US biochemists; with <u>Har Gobind Khorana</u>, US molecular chemist, approached the problem of the 'code dictionary' by synthesizing a nucleic acid with a known base sequence, found which amino acid is converted to protein; and achieved *deciphering the full code.*

1968-72: <u>Thomas Albert Sebeok</u>, Hungarian-born US semiotician and linguist, based on the theories of JJ von Uexkül, he produced works such as *Zoosemiotics*, American Speech, vol.43, no.2, and *Perspectives in Zoosemiotics*, Janua Linguarum, Seies Minor 122, The Hague: Mouton de Gruyter.

1969: <u>David Lewis</u>, US philosopher, studied signalling games, and wrote the work entitled *Convention. A Philosophical Study*, edited at Harvard University Press.

1970: <u>David Baltimore</u>, US microbiologist, discovered the *reverse transcriptase enzyme* which can transcribe DNA into RNA, allowing scientists to manipulate the genetic code.

1970-77: <u>Roger Guillemin</u>, US physiologist, isolated and identified the chemical structures of *hypothalamic hormones*, especially the hormone that stimulates the thyroid gland, and hormones releasing and inhibiting growth hormone.

1973: <u>Harry Jerison</u>, US psychiatrist, produced the work *Evolution of the brain and intelligence*, Academic Press.

1973-74: <u>Jacques Yves Cousteau</u>, French naval officer, explorer, conservationist and scientist, studied many forms of life in the sea; pioneered marine conservation, and published works including *World without Sun*; *Octopus and Squid: The Soft Intelligence*; *Three Adventures: Galápagos, Titicaca, the Blue Holes*; *Life at the Bottom of the World*; and *The Ocean World*.

1977: <u>Thomas Cech</u>, US biochemist, discovered that *protein-free precursor RNA performs its own cleavage and splicing*, acting in the manner of an enzyme, but modifying the molecule in the process. \ <u>Phillip Allen Sharp</u>, US molecular biologist, invented a *mapping technique* used extensively in analysis of RNA molecules, leading to the discovery that *genes are split into several sections, separated by stretches of DNA* known as 'introns' which appear to carry no genetic information.

1981-99: <u>Roy John Britten</u> and <u>Eric Harris Davidson</u>, US molecular biologists, studied genomes of higher organisms, elucidated genome organization, and showed that they contain DNA strands organized into *unique, single-copy DNA sequences* (coding for single genes), *moderately repetitive DNA* (coding for gene families), and *highly repetitive sequences* which are repeated hundreds of thousands of times in the genome.

1982: Ralph Holloway and David Post, US anthropologists, studied 89 primate species identified as fossil hominids, noticed the nonlinearity of the brain-to-body relationship, and introduced the *encephalization coefficient* that is also applicable for other mammals.

1983: Kary Banks Mullis, US biochemist discovered the *polymerase chain reaction* 'PCR', allowing tiny quantities of DNA to be copied millions of times for practical analysis.

1985-89: Michael Bishop, US molecular biologist and virologist, discovered *oncogenes*, normal cellular genes involved in growth and development of all mammalian cells.

1989-2000: Paul Greengard, US biochemist, Arvid Carlsson, Swedish pharmacologist, and Eric Kandel, US biologist, researched *signal transduction in the nervous system*, enabling the development of a range of new drugs.

1990-2005: Roger David Kornberg, US biochemist, studied the molecular basis of eukaryotic *transcription*, as process of making a RNA-copy of part of the DNA, followed by movement of RNA out of the cell nucleus to ribosomes.

1992-2004: Didier Raoult, Stéphane Audic, Catherine Robert, Chantal Abergel, Patricia Renesto, Hiroyuki Ogata, Bernard La Scola, Marie Suzan, and Jean-Michel Claverie, French researchers, discovered and studied *Mimivirus*, and published their well-known *1.2-Megabase Genome Sequence of Mimivirus*, Science, 306.

1994: Gordon Shepherd, US neuroscientist, worked on vertebrate neurology, and published the influential book *Neurobiology*, Oxford University Press.

1995-2010: Brian Skyrms, US researcher, made several *sender-receiver models* within the teleosemantic framework, and published the results of his work as *Signals: Evolution, Learning, and Information*, Oxford University Press.

1996: Rudolf Raff, US evolutionary biologist, wrote the book entitled *The Shape of Life: Genes, Development, and the Evolution of Animal Form*, University of Chicago Press. \ Edward Michael De Robertis, US embryologist, and Yoshiki Sasai, Japanese evolutionary biologist, researched for a common ancestor of all bilaterians, called by them *Urbilateria*, and produced a well-known work entitled *A common plan for dorsoventral patterning in Bilateria*, Nature, 380.

1999: Samuel Bowring and Ian Williams, Canadian researchers, investigated Priscoan rocks, and published their results as *Priscoan (4.00-4.03 Ga) orthogneisses from north-western Canada*, Contributions to Mineralogy and Petrology, 134.

2000-01: Simon Wilde, Australian, John Valley, US, William Peck, UK, and Colin Graham, US, researchers, studied detrital zircon in different areas, and published their findings as *Evidence from detrital zircons for existence of continental crust and oceans on the Earth 4.4 Gyr ago*, Nature Geoscience, 409.

2001: Cristopher Fedo, US, Keith Sircombe, Australian, and Robert Rainbird, Canadian, researchers, synthesized results from identification of detrital zircon crystals in sedimentary rocks, and published *Detrital Zircon Analysis of the Sedimentary Record 4.4 Ga old.*

2002: Bruce Alberts, Alexander Johnson, Julian Lewis, Martin Raff, Keith Robert and Peter Walter, US researchers into cell communication, published their work *General Principles of Cell Communication*, Molecular biology of the cell, New York, 4th ed. \ Aivo Lepland, Estonian, Gustav Arrhenius, US, and David Cornell, Swedish, researchers, studied the possibility of using graphitic rocks as potential biomarkers, and published *Apatite in Early Archean Isua Supracrustal rocks, southern West Greenland - its origin, association with graphite and potential as a biomarker*, Precambrian Research, 118. \ Stefan Bengtson, Swedish palaeozoologist, wrote *Origins and early evolution of predation*, in M Kowalewski and PH Kelley, eds. 'The fossil record of predation', The Paleontological Society Papers, 8.

2003: Bernard La Scola, Stéphane Audic, Catherine Robert, Liang Jungang, Xavier de Lamballerie, Michel Drancourt, Richard Birtles, Jean-Michel Claverie, and Didier Raoult, French researchers, worked to find out more about the newly discovered *Mimivirus*, and published their results as *A Giant Virus in Amoebae*, Science, 299. \ Alain Ghysen, researcher at the Laboratory of Neurogenetics of Montpellier University in France, presented his work *The origin and evolution of the nervous system*, International Journal of Developmental Biology, 47.

2004: Ralph Edward Hoffman, Michelle Hampson, Maxine Veranko, and Thomas McGlashan, US neuro-scientists, modelled the rearrangement in the human brain evolution, and published their work entitled *Auditory hallucinations, network connectivity, and schizophrenia*, Behavioural and Brain Sciences, 27(6).

2005: Sten Grillner, Jeanette Hellgren, Ariane Ménard, Kazuya Saitoh and Martin Wikström, Swedish neuroscientists, wrote the article *Mechanisms for selection of basic motor programs - roles for the striatum and pallidum*, Trends Neuroscience, 28(7). \ Eva Jablonka and Marion Lamb, Israeli geneticists, published their work *Evolution in four dimensions: Genetic, epigenetic, behavioural, and symbolic*, MIT Press. \ Robert Lichtneckert and Heinrich Reichert, Swiss geneticists and developmental biologists, made known their work *Insights into the urbilaterian brain: conserved genetic patterning mechanisms in insect and vertebrate brain development*, Heredity, 94.

2005-10: Aleksandra Mloszewska and Kurt Konhauser, Canadian researchers, carried out studies on the biochemical origin of old banded iron layers which were formed in sea water as the result of oxygen released by photosynthetic cyanobacteria (blue-green algae), and published *Geochemistry of > 3.8 Ga banded iron formation - Isua Supracrustal Belt, Nuvvuagittuq Supracrustal Belt - implications for the chemistry of Earth's earliest oceans*, GeoCanada, Working with the Earth.

2006: Christiane Nüsslein-Volhard, German biologist, became well-known for her studies on genetics, and wrote *Coming to Life: How Genes Drive Development*, Kales Press.

2007: Alexandru Denes, Gáspár Jékeley, Detlev Arendt, *et al.*, researchers at the European Molecular Biology Laboratory in Heidelberg, Germany, published their paper entitled *Molecular Architecture of Annelid Nerve Chord Supports Common Origin of Nervous System Centralization in Bilateria*, Cell, 129(2).

2008: Sara Mitri, Dario Floreano and Laurent Keller, researchers at the Laboratory of Intelligent Systems, EPFL, at the University of Lausanne, Switzerland, conducted experimental evolution of communication using robots, and produced the work *Evolutionary Conditions for the Emergence of Communication*, within Future and Emerging Technologies Program (IST-FET') of the EC under the EU R & D contract IST-2003. \ Leslie Pray, US geneticist, studied the eukaryotic nuclear genome, and published her paper *Eukaryotic Genome Complexity*, Nature Education, 1.

2009: John Stevely and Don Sweat, US biologists, conducted long-term research on Florida's sponge populations, and wrote *Florida's Marine Sponges: Exploring the Potential and Protecting the Resource*, Florida Sea Grant College Program, SGEF-169.

2010-11: Sun Kwok and Yong Zhang, Hong Kong scientists, researched different phases of stellar evolution, proto-planetary and planetary nebulae, and especially cosmic dust, as well as analyzed data from the European Space Agency's Infrared Space Observatory and from NASA's Spitzer Space Telescope; they found that *organic compounds commonly exist throughout the universe*, as proved by an infrared emission band indicating an amorphous carbonaceous solid with *mixed aromatic and aliphatic structures* naturally and rapidly created by stars; the results and consequences of these findings being published in their *Organic Matter in the Universe*.

2011: Timo Maran, Estonian, Dario Martinelli, Finnish, and Alexkei Turovsky, Estonian, editors, made public the book *Reading in Zoosemiotics*, Semiotics, Communication and Cognition, 8, Berlin: Mouton de Gruyter. \ Susannah Poter, US palaeobiologist, published her article *The rise of predators*, The Geological Society of America, Geology, 39(6).

2011-12: Scott Sanford, US Space science researcher at NASA Ames, studied the chemical processes that occur when high-energy ultraviolet radiation bombards simple ices like those seen in space, and found out that a surprisingly rich mixture of organics is made, such as *amino acids, nucleobases* and *amphiphiles*; his findings were published in a series of articles including *The Power of Sample Return Missions - Stardust and Hayabusa*, Proceedings of IAU Symposium #280 - 'The Molecular Universe', eds. J Cernicharo and R Bachiller.

2012: <u>Günther Witzany</u>, Austrian philosopher, and <u>Baluška František</u>, German molecular biologist, edited the book *Biocommunication of Plants*, Springer Verlag.

2013: <u>Steven Benner</u>, US geochemist and Professor at the Westheimer Institute for Science and Technology, interpreted the results of the first analyses of Martian soil by *Curiosity rover*, and said 'The evidence seems to be building that we are actually Martians; that life started on Mars and came to Earth on a rock'.

2013-14: <u>Marius Albu</u>, Romanian-born British interdisciplinary researcher, wrote the series of books *Conventional and Non-conventional Forms of Energy* (including cosmic, terrestrial, biological, human, cerebral and socioeconomic forms), *Integrated Course of Life, Soul and Mind* (displaying spatial scales in search for organic matter and sequences of evolution), both published by United p.c., European Union, and *Time, The Harmonizer of All Things* (deducing the time function from the laws of thermodynamics and applying it to various natural processes), published by Charleston SC, USA.

2014: <u>Günther Witzany</u>, Austrian philosopher, edited the work entitled *Biocommunication of Animals*, Dortrecht, Springer Verlag, revealing a systematic investigation of animal communication, a broad variety of examples for different species, and a trans-disciplinary concept of biocommunication.

2. Primates and their communication

2.1. Rise and diversification of primates

The scientific study of both living and extinct primates with their evolution and behaviour in natural habitats and in laboratories is called _primatology_. It is closely related to anatomy, palaeontology, anthropology, biology, zoology, veterinary sciences, medicine, genetics, neurology, psychology, sociobiology, and recently dating techniques.

Primatology is mainly focused on the common characteristics of primates and humans in order to understand the links in their evolution. In taxonomy (JR Napier and P Napier, 1985; CP Groves, 2001), primates are divided into two main groupings: _prosimians_ (lemurs, lorisoids, and tarsiens), and _simians_ (monkeys, apes, and hominids).

Simians are subdivided into two groups: _Catarrhini_ 'narrow-nosed monkeys', consisting of Old World monkey (baboons and macaques), gibbons, and great apes; and _New World monkeys_ 'flat-nosed monkeys', including the capuchin, howler, and squirrel monkeys. Recently, taxonomists have preferred to split primates into the suborders _Strepsirrhini_ 'wet-nosed primates', and _Haplorhini_ 'dry-nosed primates'. Meanwhile, primatology tends to be more associated with sociobiology, in order to identify the roots of our social behaviour. Modern molecular biology reinforced humanity's place within the Primate order. Humans and simians share the vast majority of their DNA, for example the chimpanzees share between 97% and 99% genetic identity with humans.

Primatology was developed by studies and experiments differently accomplished by two 'schools' of primatology, namely the Western primatology (in Europe and North America), and East Asian primatology (in Japan and China).

Western primatology was approached in three methodological ways: _(i)_ field study, as the more realistic; _(ii)_ laboratory study, as the more controlled; and _(iii)_ semi-free ranging, where primate habitat and wild social structure is replicated in a captive setting (J Goodall, 1969-2009; D Fossey, 1974-83).

Japanese primatology was developed out of animal ecology, tending to be more interested in the social aspect of primates, and looking for an insight into the duality of human nature, as individual self versus social self. This East Asian primatology is based on data such as: _(i)_ data coming from identification with the studied subject, _e.g._ by domestication; and *(ii)* data from the behaviour of an individual in a group for unveiling their complex interaction (J Itani, 1965-94; K Imanishi, 1992-2002).

Primates arouse from ancestors living in the trees of tropical forests, and still many of their characteristics represent adaptations to life in such a challenging environment, and most primate species remain at least partly arboreal. They range in size from *Madame Berthe's mouth lemur*, which weighs only 30 grams, to *eastern lowland gorilla*, weighing over 200 kilograms.

The earliest placental mammals, such as *Eutheria* (TH Huxley, 1863-80) represented by the extinct genus *Juramaia sinensis*, were found in Liaoning, China, and dated as around 160 million year old. During a major glacial period between 150 and 130 million years ago, when the average global temperature reached its minimum of only 16°C about 142 million years ago, the environmental conditions on the Earth gradually improved, and thus the mammal evolution accelerated and diversified, leading to the appearance of first primates around 100 million years ago. On the primate evolutionary line, the oldest known primate-like mammal species, called *Plesiadapis*, lived 58-55 million years ago and was spread in North America, Eurasia and Africa during the tropical conditions of the early Cenozoic.

Molecular clock studies suggest that the primate branch may be originated around 85 million years ago; while fossil evidences of the earliest known true primate, namely the genus *Teilhardina*, is dated as about 56 million year old.

According to recent researches, crucial genetic differences have emerged in primate evolution when new forms appeared, in million years ago, as: *Lemuriformes*, 85-65; *Tarsiiformes*, 60-50; *Platyrrhini*, 45-35; *Cercopithecinae*, around 25; *Hylobatidae*, about 18; *Pongidae*, about 14.

It is thought that in primates' lineage there was an ancestor of all species related to *Aegyptopithecus*, *Propliopithecus* and *Parapithecus*, which have been unearthed in the area of Fayum, Egypt, and dated around 35 million years old. Subsequently, *Saadanius* was described as a close relative of the last common ancestor of the *crown catarrhines* from 29-28 million years ago.

The earliest known catarrhine was *Kamoyapithecus* found in upper Oligocene at Eragaleit in the Northern Kenya Rift Valley, and dated c.25 million years ago. Fragments of a fossil *Victoriapithecus*, the earliest of Old World Monkeys, were dated 20 million years ago. The series consisting of *Proconsul*, *Dendropithecus*, *Afropithecus*,

57

Heliopithecus, *Kenyapithecus*, *Rangwapithecus*, *Limnopithecus*, *Nacholapithecus*, *Equatorius*, and *Nyanzapithecus* were identified in East Africa and dated between 23 and 12 million years ago.

The lineage of *gibbons* diverged from Great Apes between 18 and 12 million years ago; and *orang-utans* diverged from the other Great Apes around 12 million years ago.

In succession, the fossil *Griphopithecus*, in Turkey, and proto-orangutan *Sivapithecus*, in India, are dated as 15-9 million year old.

Descending from earlier primates, the African apes developed by other forms, including the 11.5 million year old genus of apes called *Dryopithecus*.

Studies carried out on the Eocene-Oligocene fossil beds of Fayum depression, southwest of Cairo, gave a wider view on the evolution of tropical population of primates which leads to the present-day species belonging to *lemurs* in Madagascar; *lorises* in Southeast Asia; *galagos* 'bush-babies' in Africa; and *anthropoids* such as *platyrrhine* 'New World monkeys', *catarrhines* 'Old World monkeys', and *great apes*, including humans.

Today, the order of Primates consists of more than 300 species, being the third most diverse order of mammals, after rodents and bats.

2.2. Adaptive changes

Primates are distinguished by their *anatomical, physiological and morphological characteristics* commonly presented as: *binocular vision*, allowing accurate distance perception; *large domed cranium*, particularly in anthropoids; *elaborated brain*, especially the neocortex involving sensory perception, generation of motor commands, and spatial reasoning conscious thought; as well as *opposite thumbs* of many higher primates, allowing them to use tools (DJ Morris, 1965-94; F de Waal, 1989-2013). Their *brain growth* appear to bear at least two major metabolic adaptations: during the prenatal period, showing heavy maternal investment; and in course of postnatal period, implying a special mother's protection.

Three-colour vision and opposable thumbs are distinctive for many primates, which also have slower rates of growth than other similarly sized mammals and reach maturity later, but have longer lifespans.

Whereas all other mammals have claws or hooves on their digits, only primates have flat nails. Some primates have claws, but even among these there is a flat nail on the big toe-hallux. In all primates, excepting humans, the hallux diverges from the other toes and together with them forms a pincer for grasping objects such as branches. Although not all primates have similarly dextrous hands, the catarrhines (Old World monkeys, apes and humans) and some lemurs and lorises have an opposite thumb.

Primates, like other arboreal mammals (e.g. squirrels and opossums), possess grasping feet and most of them are arboreal, so that they maybe evolved from an arboreal ancestor. The possession of specialized nerve endings represented by tactile corpuscles, called *Meissner's corpuscles*, which are a sort of mechanoreceptors situated in nerve terminals of the skin in hands and feet, give to primates an increasing tactile sensitivity (G Meissner, 1852-53). It seems that no other placental mammals have such tactile corpuscles.

In all primates the eyes face forward and then the eyes' visual fields are overlapping. This feature is not restricted to primates, but it is generally common among the predators, and therefore the primates' ancestor could have also been a predator, perhaps insectivorous.

The optic fibres in almost all mammals cross over (decussate) so that signals from one eye are interpreted on the opposite side of the brain, but in some primate species up to 40% of the nerve fibres do not cross over.

Primate teeth are distinctive from those of other mammals by the low, rounded form of the molar and premolar cusps, contrasting with the high, pointed cusps or elaborate ridges of other placental mammals.

Though primates began as an arboreal group, and many of them remained arboreal, others became at least partly terrestrial and further have achieved higher levels of intelligence. Many primates sit upright without supporting the body weight by their arms, while the apes can walk upright for short distances only.

Comparing to other mammals, primates have larger _brains_ and an increased reliance on stereoscopic vision at the expense of smell that is the dominant sensory system in most mammals. These features are more developed in monkeys and apes, and noticeably less in lorises and lemurs. Most of the enlargement of the primate brain comes from a massive expansion of the cerebral cortex, especially the _prefrontal cortex_ and the parts of the cortex involved in vision. It has been estimated that visual processing areas occupy more than half of the total surface of the primate _neocortex_. The prefrontal neocortex carries out functions that include planning, working memory, motivation, attention, and executive control, so that it takes up a much larger proportion of the brain for primates than for other species, and an especially large fraction of the human brain.

When the primate brain increases in size, not all parts increase at the same rate, and thus the larger brain has a greater fraction taken up by the neocortex. Pre-hominid primates have cranial capacities such as:

Primates	Cranial capacity (cm^3)
Tarsiers	3.5 - 3.7
Lorises	3 - 14
Lemurs	9 - 23
Squirrel monkeys	24 -25
Gibbons	82 - 104
Rhesus macaques	100 - 110
Siamangs	125 - 130
Bonobos	179 - 336
Orang-utans	275 - 443
Chimpanzees	321 - 500
Gorillas	340 - 752

Cortical areas for vocalization and later for speech, such as _anterior circulate cortex, cingulate gyrus, Broca's area_ and _Wernicke's area_ are present especially in great apes, e.g. gorillas have the anterior circulate cortex and cingulate gyrus, while chimpanzees have in addition both Broca's and Wernicke's areas, as homologous to cerebral areas of the humans. Remarkably, Broca's area is located in the posterior portion of the frontal lobe (P Broca, 1861-78), and Wernicke's area is located in the posterior portion of the left temporal lobe (C Wernicke, 1874-85), i.e. both in the left hemisphere of the brain, where they are connected by a large bundle of nerve fibres called *arcuate fasciculus*.

Compared with body weight, the primate brain is larger than that of other terrestrial mammals, and it has a fissure unique to primates, namely the *Calcarine sulcus*, that separates the first and second visual areas on each side of the brain.

Neuroanatomical changes relevant to speech that accompanied divergence from the last common ancestor of chimpanzees, bonobos and humans are associated with cytoarchitectonic boundaries of the so-called *temporoparietal* (Tpt) *area*, a component of Wernicke's area, in common chimpanzee brains, which was studied by stereological methods to estimate regional volumes, total neuron number and neuron density (M Spocter *et al.*, 2010). This study unveiled that chimpanzees display significant population-level leftward asymmetry of area Tpt in terms of neuron number, with volume asymmetry approaching significance; and furthermore, asymmetry in the number of neurons in the same area was positively correlated with asymmetry of neuron numbers in Brodmann's area 45 (K Brodmann, 1909), a component of Broca's frontal language region. In conclusion, the leftward asymmetry of Wernicke's area originated prior to the appearance of modern human language and *before* our divergence from the last common ancestor.

Adult primates may live in solitude or in mated pairs, but most of them live in groups up to hundreds of members, developing social behaviours. Among the higher mammals, most *primates* spend their lives in complex, tightly woven societies (B Smuts *et al.*, 1987) and frequently need to communicate with each other, such as by smells, sounds, gestures, visual messages, touching, and scent marking (way of claiming territory and warning off intruders, such as by urine, as a personal signature). Such kinds of communication are displayed by *gorillas* when stick out their tongues to show anger, and by *chimpanzees* when greet each other by touching hands.

Researches with chimpanzees, bonobos, and gorillas indicate their relatively complex systems of communication. At least the *African apes* can learn and use a simplified version of the American Sign Language (ASL) for the Deaf.

Many primates produce male and female *pheromones* which play a part in sexual attraction and ovulation regulation.

The dominant male in a monkey or ape community can prevent major conflicts and keep order by use of quite subtle *agonistic displays*. For instance, male baboons flash their eyelids when they are angry and want to intimidate others; in the case that it is not sufficient in its effect, they open their mouths widely in a manner looking like human yawning, as the last warning before attacking.

Most primate species communicate affection and reduce group tension by what are known as *affiliative behaviours*. At a basic level, various primates produce *utterances*, consisting of combinations of calls, which can be meaningful to others. By difference from human communication, primate sounds are reflexive, genetically determined, and affective (emotional), while their vocalizations are quite voluntary and representational.

Primate *social behaviours* were recently researched on one of our more genetically distant primate relatives, the cotton-top tamarin (*Saguinus oedipus*), that is a small New World monkey, living in Columbia and weighing less than 0.5 kilogram, easily recognizable by the long white sagittal crest extending from its forehead to its shoulders. It displays a wide variety of social behaviours, forming dominance hierarchy groups in which only dominant pairs breed. The tamarins have high level of cooperative care, as well as altruistic and spiteful behaviours. *Communication* between them is sophisticated and shows evidence of grammatical structure, a language feature that must be acquired. They vocalize with birdlike whistles, soft chirping sounds, high-pitched trilling, and staccato calls. There were detected 38 distinct sounds as unusually sophisticated, conforming to grammatical rules and using both phonetic and lexical syntax (J Cleveland and C Snowdon, 1982). By this range of vocalizations, the adults may be able to communicate with one another about intention, thought processes, and emotion including curiosity, fear, dismay, playfulness, warning, joy, and calls to young. Then the study was directed to tamarin infants, showing that they can respond behaviourally to vocalizing adults in a fashion indicating their comprehensibility of auditory inputs (N Castro and C Snowdon, 2000). The further studies showed that the cotton-top tamarins can learn to generalize algebraic rules, which is also a skill that human infants use in acquiring language. If this case, it can be posited that some of the cognitive processes underlying language probably existed in the most distant human/primate ancestors (M Hauser *et al.*, 2009).

In general, primate *cognition and communication* are referring to:
- Making and using *tools* to acquire food and for social display;

- Sophisticated hunting strategies requiring *cooperation, influence and rank*;
- *Olfactory signals* for many aspects of social and reproductive behaviour in the case of lemurs, lorises, tarsiers and New World monkeys, and well as specialized *glands used to mark territories* with pheromones as a process forming a large part of the common behaviour of these primates;
- High auditory sensibility for *ultrasonic communication*, such as some Philippine tarsiers whose ultrasonic vocalization represents a private channel of communication that subverts detection by predators, pray and competitors.

Sign language of apes in the wild is gestural communication with a repertoire of at least 66 different gestures, including bodily movements, which comprise almost all types of gesture for chimpanzees; and also 30 different manual gesture types in mature chimpanzees, such as arm beckon, point, clap, and flail. They can learn up to 125 signs in 44 months, but their communication is symbolic, and lacks grammar or rules. However, the apes fail to ask questions themselves, this cognitive ability being characteristic only for humans who from childhood ask interminable questions.

Primates make 'vocal calls' which are generated by circuits in the brainstem and limbic system. The vervet monkeys are able to make up to ten different *vocalizations*, many of them being used to warn other members of the group about approaching predators, and including a 'leopard call', a 'snake call', and an 'eagle call'. Their vocalizations are also used for identification, for example if an infant monkey calls, its mother turns toward it; while other vervet monkeys turn instead toward that infant's mother to see what she will do.

Gestural language depends on similar neural systems as the vocal language, i.e. mostly in the brain's left-hemisphere, on the cortical areas responsible for mouth and hand movements of higher primates which can use gestures or symbols for primitive forms of communication, some of them resembling those of humans, such as 'begging posture'.

Great ape communication in the wild is governed by cerebral centres in Broca's and Wernicke's areas, which are responsible for: controlling the *muscles* of the face, tongue, mouth, larynx, and recognizing sounds; enabling them to make *vocal calls* generated by circuits in the brainstem and limbic system; as well as *chattering* controlled by Broca's area, and *vocalization*, such as chimpanzees, using the same brain regions as humans hearing speech.

Research into great ape language has involved teaching chimpanzees, bonobos, gorillas and orang-utans to communicate with human beings and with each other using sign language, physical tokens and lexigrams. Based on these means of communication, some primatologists argue that primates have the ability to communicate apparently by 'language', although the non-consistency with this term, because any language has a grammar and a set of lexemes.

Apes produce vocalizations in presence of emotional states, and much less or at all in the absence of such states.

The Broca's and Wernicke's areas in the primates' brains are responsible for controlling the muscles of face, tongue, mouth, larynx, and also for recognizing sounds. Primates' vocal calls are generated by circuits in the brainstem and limbic system.

The use of 'syntax' in non-humans, and the ability to produce 'sentences' were prefigured so early as displayed by *Cercopithecus nictitans* (the greater spot-nosed monkeys) of Nigeria which can take discrete units of communication and build them up into a sequence that carries a different meaning from the individual 'words'. These monkeys have two main alarm sounds: *onomatopoeic* as the 'pyow' warning against a lurking leopard, and coughing sounds believed to be a 'hack' used when an eagle is flying nearby (The Times, May 18, 2006). Similar conclusions were reported in the case of *Campbell's Mona monkeys*, which also use calls with specific meanings: 'hok' for eagle, 'krak' for leopard, 'krak-oo' for general disturbance in forest canopies, 'boom' for non-predators when calling a group together to travel or arguing with neighbouring groups (B Keim, 2009-13). Primate communication can also be evidenced through the use of *lexigrams*, such as by bonobos and chimpanzees, in addition to their signs.

Chimpanzees produce vocalizations by alternating the sizes and shapes of their mouths and resonating cavities, and 'facial expressions' play a key role in close-up communication between them. There were discovered 34 discrete calls along with the emotions with which they are associated. Chimpanzees' vocalizations include *hoos, screams, whimpering, tantrum screems*, and also *effort-grunts, staccatos* and *uhgrunts* (Dame JM Goodall, 1986).

Modularity of the primate mind was studied on orang-utans which are currently being taught language at the Smithsonian National Zoo, in Washington DC, USA, using a computer system, and indicating that human babies and grown monkeys approach and process numbers in a similar fashion (F Neago, 2006).

Some modern psychologists have the idea that primate grooming is a social activity that strengthens relationships, and even suggest a link between primate grooming and the development of speech, in terms of 'Monkeys and apes are just as social as we are, just as intensely interested in the social whirl around them'; 'primates live in very much more sophisticated kind of social world than other animals do', and 'language was not invented for communication but as a more efficient form of grooming, which in turn functioned to cement society' (R Dunbar, 2011).

2.3. Timeline sequencing

Primate evolution (J Fleagle, 1999) was marked by a series of genera and species, including those mentioned in the previous two subchapters from which there were selected the following ones dating, in million years ago, such as:

58 - 55: *Plesiadapis*, a primate-like mammal that lived in North America and Europe (FLP Gervaise, 1877);

56 - 47: *Teilhardina*, an early marmoset-like primate that was widely spread in Europe, North America and Asia (GG Simpson, 1940);

35 - 33: *Aegyptopithecus*, an early fossil catarrhine in Egypt (ES Simons, 1965);

33: *Parapithecus*, an extinct genus of primate that lived in Egypt (M Schlosser, 1911; ES Simons, 1965; KC Beard, 2002);

30 - 25: *Propliopithecus*, an extinct genus of ape that lived in Egypt (D Palmer, 1999; ES Simons *et al.*, 2005);

29 - 28: *Saadanius hijazensis*, crown catarrhines which lived in Saudi Arabia (I Zalmout *et al.*, 2009-10);

27.5 - 24.2: *Kamoyapithecus*, a large primate that lived in Kenya (REF Leakey *et al.*, 1995);

23 - 14: *Proconsul africanus*, an extinct primate with fossils in Eastern Africa (AT Hopwood, 1933; LSB Leakey and M Leakey, 1934-50);

20: *Victoriapithecus*, an early Old World monkey skull, discovered near Lake Victoria in Kenya (B Benefit and M McCrossin, 1990-91);

20 - 15: *Dendropithecus*, an extinct ape genus native to East Africa (P Andrews and ES Simons, 1977);

18 - 16: *Afropithecus*, a hominoid excavated near Lake Turkana in Kenya (REF Leakey *et al.*, 1986-88);

16: *Heliopithecus*, fossil remains of a jaw and teeth found in Saudi Arabia (P Andrews and L Martin, 1987);

15: *Griphopithecus*, a prehistoric ape that lived in Turkey and Central Europe (T King *et al.*, 1999);

14: *Kenyapithecus*, an upper jaw and teeth from a fossil ape found at Fort Ternan in Kenya (LSB Leakey, 1961);

14 - 12: *Rangwapithecus*, an extinct genus of ape that lived in Kenya (A Hill, *et al.*, 1987);

12.5 - 8.5: *Sivapithecus*, remains of the face and jaw from an extinct primate that lived in the Indian subcontinent (D Pilbeam, 1982);

11.5: *Dryopithecus*, fossil genus of ape that lived in Eastern Africa and Eurasia (L Kordos and D Begun, 2001);

9 - 7: *Oreopithecus*, fossil genus with the representative bipedal-like species *Ouranopithecus bambolii* that lived in Tusco-Sardinian area, Italy (FLP Gervaise, 1872; J Hürzeler, 1949-60);

8 - 4: *Early hominids* split first from gorillas and then from chimpanzees (so that the humans and chimpanzees have 98.4% of DNA common).

At full scale of primate evolution from the placental mammals to hominids and from hominids to humans, a preliminary timeline can be drawn up by sequences with transitional times, in million years ago, as follows:

\ 84 \ Placental mammals

\ 79 \ Proto-primates

\ 73 \ Lemuriformes

\ 68 \ early primates

\ 63 \ Tarsiiformes

\ 58 \ Plesiadapis, Teilhardina

\ 53 \ Tarsiers

\ 48 \ Simians

\ 43 \ New World monkeys

\ 39 \ Aegyptopithecus

\ 34 \ Parapithecus

\ 30 \ Propliopithecus, Saadanius

\ 25 \ Kamoyapithecus, Proconsul

\ 21 \ Victoriapithecus, Dendropithecus, Afropithecus

\ 16 \ Heliopithecus, Griphopithecus, Kenyapithecus, Rangwapithecus

\ 12 \ Sivapithecus, Dryopithecus, Oreopithecus

\ 8 \ early hominids

\ 4 \ late hominids

\ 0.2 \ Humans emerged.

The sequences above can be integrated in the evolution of life by recounting the transitional times in million years ago as times from the origin of life c.4200 million years ago, i.e.

4200 - 84 = 4116 \ 4200 - 79 = 4121 \ 4200 - 73 = 4127 \
4200 - 68 = 4132 \ 4200 - 63 = 4137 \ 4200 - 58 = 4142 \
4200 - 53 = 4147 \ 4200 - 48 = 4152 \ 4200 - 43 = 4157 \
4200 - 39 = 4161 \ 4200 - 34 = 4166 \ 4200 - 30 = 4170 \
4200 - 25 = 4175 \ 4200 - 21 = 4179 \ 4200 - 16 = 4184 \
4200 - 12 = 4188 \ 4200 - 8 = 4192 \ 4200 - 8 = 4192 \
4200 - 4 = 4196 \ 4200 - 0.2 = 4199.8 \ the present time.

In the timescale of evolutionary life (see table at the end of subchapter 1.3) the most recent $N/10 = \alpha$ sequences are delimited by values of argument \1.7\ ... \1.8\ ... \1.9\ corresponding to the entire evolution of primates. In order to detail primates' evolution, these most recent sequences are subdivided into $N/10^2 = \beta$ sequences as suitable for a timeline marking the succession of selected representatives from the placental mammals through hominids to humans. According to the final time $t_\bullet$ = 4400 million years from the origin of life, primates sequences are delimited by the transitional times $t_\beta = t_\bullet \cdot tanh(\beta)$ calculated, in million years from the same origin, such as:

$$t_{1.70} = t_\bullet \cdot tanh(1.70) \approx (4400) \cdot (0.9354091) \approx 4115.80;$$
$$t_{1.71} = t_\bullet \cdot tanh(1.71) \approx (4400) \cdot (0.9366475) \approx 4121.25;$$
$$t_{1.72} = t_\bullet \cdot tanh(1.72) \approx (4400) \cdot (0.9378630) \approx 4126.60;$$
$$t_{1.73} = t_\bullet \cdot tanh(1.73) \approx (4400) \cdot (0.9390559) \approx 4131.85;$$
$$t_{1.74} = t_\bullet \cdot tanh(1.74) \approx (4400) \cdot (0.9402266) \approx 4137.00;$$
$$t_{1.75} = t_\bullet \cdot tanh(1.75) \approx (4400) \cdot (0.9413755) \approx 4142.05;$$
$$t_{1.76} = t_\bullet \cdot tanh(1.76) \approx (4400) \cdot (0.9425030) \approx 4147.01;$$
$$t_{1.77} = t_\bullet \cdot tanh(1.77) \approx (4400) \cdot (0.9436094) \approx 4151.88;$$
$$t_{1.78} = t_\bullet \cdot tanh(1.78) \approx (4400) \cdot (0.9446952) \approx 4156.66;$$
$$t_{1.79} = t_\bullet \cdot tanh(1.79) \approx (4400) \cdot (0.9457606) \approx 4161.35;$$
$$t_{1.80} = t_\bullet \cdot tanh(1.80) \approx (4400) \cdot (0.9468060) \approx 4165.95;$$
$$t_{1.81} = t_\bullet \cdot tanh(1.81) \approx (4400) \cdot (0.9478318) \approx 4170.46;$$
$$t_{1.82} = t_\bullet \cdot tanh(1.82) \approx (4400) \cdot (0.9488384) \approx 4174.89;$$
$$t_{1.83} = t_\bullet \cdot tanh(1.83) \approx (4400) \cdot (0.9498261) \approx 4179.23;$$
$$t_{1.84} = t_\bullet \cdot tanh(1.84) \approx (4400) \cdot (0.9507951) \approx 4183.50;$$
$$t_{1.85} = t_\bullet \cdot tanh(1.85) \approx (4400) \cdot (0.9517460) \approx 4187.68;$$
$$t_{1.86} = t_\bullet \cdot tanh(1.86) \approx (4400) \cdot (0.9526788) \approx 4191.79;$$
$$t_{1.87} = t_\bullet \cdot tanh(1.87) \approx (4400) \cdot (0.9535941) \approx 4195.81;$$
$$t_{1.88} = t_\bullet \cdot tanh(1.88) \approx (4400) \cdot (0.9544921) \approx 4199.77;$$
$$t_{1.89} = t_\bullet \cdot tanh(1.89) \approx (4400) \cdot (0.9553731) \approx 4203.64.$$

By these transitional times, the sequences of primate evolution are integrated in the general life evolution and communication development, as shown in the table below.

$N/10^2$ $= \beta$	Time (million years)		Sequences of primate evolution
	from origin $t_\beta = t_\bullet \cdot tanh(\beta)$	from present $t_\beta - 4200$	
	4400	+200	
...	...	...	
1.89	4203.64	+3.64	_
1.88	4199.77	-0.23	_Humans
1.87	4195.81	-4.19	_Late hominids
1.86	4191.79	-8.21	_Early hominids
1.85	4187.68	-12.32	_Siva-, Dryo-, Oreopithecus
1.84	4183.50	-16.50	_Helio-, Gripho-, Kenya-, Rangwapithecus
1.83	4179.23	-20.77	_Victoria-, Dendro-, Afropithecus
1.82	4174.89	-25.11	_Kamoyapithecus, Proconsul
1.81	4170.46	-29.54	_Propliopithecus, Saadanius
1.80	4165.95	-34.05	_Parapithecus
1.79	4161.35	-38.65	_Aegyptopithecus
1.78	4156.66	-43.34	_New World monkeys
1.77	4151.88	-48.12	_Simians
1.76	4147.01	-52.99	_Tarsiers
1.75	4142.05	-57.95	_Plesiadapis, Teilhardina
1.74	4137.00	-63.00	_Tarsiiformes
1.73	4131.85	-68.15	_Early primates
1.72	4126.60	-73.40	_Lemuriformes
1.71	4121.25	-78.75	_Proto-primates
1.70	4115.80	-84.20	_Placental mammals
...	...	...	
0	0	-4200	

References

1852-53: <u>Georg Meissner</u>, German anatomist and physiologist, first described the tactile corpuscles, called today *Meissner's corpuscles*, and published his work entitled *Beitraege zur Anatomie und Physiologie der Haut* 'Contribution to the Anatomy and Physiology of the Skin', Leipzig.

1861-78: <u>Paul Broca</u>, French surgeon and anthropologist, first located the motor speech centre in the brain, since known as the *convolution of Broca* 'Broca's gyrus'; proposed the *motor aphasia*, the meaning of aphasia being the loss of speech marked by severe defect in the understanding of speech; his anthropological investigations giving strong support to CR Darwin's theory of the evolutionary descent of man.

1863-80: <u>Thomas Henry Huxley</u>, English biologist, published his work *Evidence as to Man's Place in Nature*, and introduced the term *Eutheria*, one of two mammalian clades that diverged by the mid Jurassic.

1872: <u>François Louis Paul Gervaise</u>, French-Swiss palaeontologist and entomologist, discovered and described the 'biped' hominid *Oreopithecus* in his published work *Sur un single fossile, d'espèce non encore décrite, qui a été découvert au Monte-Bamboli (Italie)*, Comptes rendus de l'Académie des sciences de Paris, LXXIV; but this specimen had a peculiar form of bipedalism, much different from that of Australopithecines.

1874-85: <u>Carl Wernicke</u>, German neurologist and psychiatrist, published *Der Aphasische Symtomencomplex* 'The Aphasic Syndrome', revealing *aphasia* 'loss of speech' as marked by a severe defect in understanding of speech, known as *sensory aphasia*, and showing that this kind of aphasia is typically localized in the so-called *Wernicke's area* of the left temporal lobe.

1877: <u>François Louis Paul Gervaise</u>, French-Swiss palaeontologist and entomologist, first discovered the fossil *Plesiadapis tricuspidens* in France.

1909: <u>Korbinian Brodmann</u>, German neurologist, became famous for his work of zoning the cerebral cortex in 52 distinct regions from their cytoarchitectonic (histological) characteristics, known as *Brodmann areas*; his original research on cortical cytoarchitectonics was published in *Vergleichende Lokalisationslehre der Großhirnrinde in ihren Prinzipien dargestellt auf Grund des Zellenbaues* 'Comparative Localization Studies in the Brain Cortex, its Fundamentals Represented on the Basis of its Cellular Architecture'.

1911: <u>Max Schlosser</u>, German scientist, analyzed a fragment of lower jaw with dentition from *Parapithecus*, found in the early Oligocene deposits of Fayum province, Egypt, and published his observations as *Beiträge zur Kenntnis der Oligozänen Landsäugetiere aus dem Fayum (Ägypten)* 'Contribution to knowledge about the Oligocene land-mammal from Fayum (Egypt)', Beiträge zur Paläontologie und Geologie, Geol. Österreich-Ungarns, 24.

1933: <u>Arthur Tindell Hopwood</u>, British palaeontologist, studied in detail the fossil remains from a primate which were previously discovered in the Eastern

Africa, and reclassified them as belonging to *Proconsul africanus*, this name being coined by him.

1934-50: <u>Louis Seymour Bazett Leakey</u>, British archaeologist, and his wife <u>Mary Leakey</u>, British paleoanthropologist, wrote the book *Adam's Ancestors: The Evolution of Man and His Culture*; amassed fossils of australopithecines, including the extinct ape *Proconsul*, at Olduvai and Lake Turkana; then extended their research in the Kenyan Rift Valley, and published another book entitled *Excavations at Njoro River Cave*.

1940: <u>George Gaylord Simpson</u>, US palaeontologist, wrote the paper *Studies on the Earliest Primates*, Bulletin of the American Museum of Natural History, 77(4).

1949-60: <u>Johannes Hürzeler</u>, Swiss palaeontologist, published his works *Neubeschreibung von Oreopithecus bambolii Gervais* 'New description of Oreopithecus bambolii Gervais', Schweizerishe Paläontologische Abhandlungen, 66; and *The significance of Oreopithecus in the genealogy of man*, Triangle, 4.

1961: <u>Louis Seymour Bazett Leakey</u>, British archaeologist, discovered remains from a fossil primate, named *Kenyapithecus*, at the site Fort Ternan in Kenya, and then published his findings as *A new Lower Pliocene fossil primate from Kenya*, The Annals and Magazine of Natural History, 4(13).

1965: <u>Elwyn Seymour Simons</u>, US primate anthropologist and anatomist, discovered *Aegyptopithecus zeuxis* in the Jebel Qatrani Formation of the Fayum province, Egypt, and published his article *New fossil apes from Egypt and the initial differentiation of Hominoidea*, Nature, 205; and also he wrote the article *The cranium of Parapithecus grangeri, an Egyptian Oligocene anthropoidean primate*, Proceedings of the National Academy of Sciences of the USA, 98(14).

1965-94: <u>Desmond John Morris</u>, English zoologist, ethologist, human sociobiologist and surrealist painter, became one of the world's greatest experts on primate behaviour, being well-known for works such as *The Mammals: A Guide to the Living Species*; *The Naked Ape*; *The Human Zoo*; and *The Human Animal*. \ <u>Junichiro Itani</u>, Japanese ecologist and anthropologist, was one of the founders of Japanese primatology, wrote *Hominid culture in primate perspective*, in D Quiatt and J Itani, eds.; and initiated studies for investigation of the wild chimpanzees of western Tanzania.

1969-2009: <u>Jane Goodall</u>, British primatologist, ethologist and anthropologist, studied the chimpanzees and other wild animals, and published works such as *My Friends the Wild Chimpanzees*, Washington DC: National Geographic Society; and *Hope for Animals and Their World: How Endangered Species Are Being Rescued from the Brink*, Grand Central Publishing.

1974-83: <u>Dian Fossey</u>, US zoologist, extensively studied a number of gorilla groups, becoming known by her research and conservation of the mountain gorilla; she published the book *Gorillas in the Mist*, Houghton Mifflin

Company, and the article *Mountain gorilla research*, National Geographic Society Research Reports, 14.

1977: Peter Andrews, British anthropologist, and Elwyn Seymour Simons, US primate anthropologist and anatomist, researched the fossil remains from *Dendropithecus*, and made known their results by the paper *A New African Miocene Gibbon-like Genus, Dendropithecus (Hominoidea, Primates) with Distinctive Postcranial Adaptations: Its Significance to Origin of Hylobatidae*, Folia Primatologica 28(3).

1982: Jayne Cleveland and Charles Snowdon, US neuroendocrinologists and primatologists, published their article *The complex vocal repertoire of the adult cotton-top tamarin (Saguinus oedipus oedipus)*, Zeitschrift für Tierpsychologie, 58(3). \ David Pilbeam, US paleoanthropologist, analyzed the remains from a primate fossil discovered in the sub-Himalayan Siwalik Hills, called *Sivapithecus indicus*, and published a description of its extant face and jaw, in the paper *New hominid skull material from the Miocene of Pakistan*, Nature, 295.

1985: John Russel Napier and Prue Napier, British primatologists, wrote the book *The Natural History of the Primates*, which was published by British Museum, London.

1986: Dame Jane Morris Goodall, British primatologist, ethologist and anthropologist, published her work *The Chimpanzees of Gombe: Patterns of Behaviour*, The Belknap Press of Harvard University Press.

1986-88: Richard Erskine Frere Leakey, Kenyan paleoanthropologist, Meave Leakey, London-born Kenyan paleoanthropologist, and Allan Walker, US anthropologist, first discovered the fossil remains from *Afropithecus* near Lake Turkana in Kenya, and then wrote the article *Morphology of Afropithecus turkanensis from Kenya*, American Journal of Physical Anthropology, 76(3).

1987: Barbara Smuts, Dorothy Cheney, Robert Seyfarth, Richard Wrangham, Thomas Struhsaker and David Hamburg, US psychologists and anthropologists, published the book entitled *Primate societies*, University of Chicago Press. \ Peter Andrews, British, and Lawrence Martin, US, anthropologists, studied the fossil primate remains in Saudi Arabia, showing that they are from a primate named by them *Heliopithecus*, and published their results as *Cladistic relationship of extant and fossil hominoids*, Journal of Human Evolution, 16(1). \ Andrew Hill, Isaiah Nengo Odhiambo, and James Rossie, US anthropologists, found and analyzed remains of the genus ape *Rangwapithecus* in Kenya, and made known their results by an article entitled *New Mandible of Rangwapithecus from Songhor, Kenya*, American Journal of Physical Anthropology, 72(210).

1989-2013: Frans de Waal (Franciscus Bernardus Maria 'Frans' de Waal), Dutch primatologist and ethologist, studied both the chimpanzee and the bonobo, two species whose DNA are nearly identical with ours, and wrote a series of books, including *Peacemaking Among Primates*; *Tree of Origin: What Primate Behaviour Can Tell Us about Human Social Evolution*; *Our*

72

Inner Ape: the Past and Future of Human Nature; and *The Bonobo and the Atheist*.

1990-91: Brenda Benefit and Monte McCrossin, US anthropologists, she discovered the fossil *Victoriapithecus macinnesi* near Lake Victoria in Kenya, and then together published the paper *Ancestral facial morphology of Old World higher primates*, Proceedings of the National Academy of Sciences, 88(12).

1992-2002: Kinji Imanishi, Japanese ecologist and anthropologist, was co-founder of Japanese primatology, studied the wild Japanese monkeys, and published his work *The World of Living Things*.

1995: Richard Erskine Frere Leakey, Kenyan paleoanthropologist, Peter Ungar, US anthropologist, and Alan Walker, English anthropologist and biologist, researched the fossil remains of *Kamoyapithecus*, and wrote the article *A new genus of large primate from the late Oligocene of Lothidok, Turkana District, Kenya*, Journal of Human Evolution, 28.

1999: John Fleagle, US anthropologist and primatologist, wrote the book entitled *Primate Adaptation and Evolution*, 2nd ed., Academic Press, New York. \ Douglas Palmer, English Earth science writer, was author of *The Marshal Illustrated Encyclopaedia of Dinosaurs and Prehistoric Animals*, London: Marshal Editions Developments Ltd., where *Propliopithecus* is presented. \ Tania King, Leslie Aiello and Peter Andrews, British anthropologists, analyzed the dental features of a species of *Griphopithecus*, and made known their results by the paper entitled *Dental microwear of Griphopithecus alpani*, Journal of Human Evolution, 36.

2000: Nicole Castro and Charles Snowdon, US neuroendocrinologists and primatologists, carried out studies on infants of cotton-top tamarins, and their findings were published in the article *Development of vocal responses in infant cotton-top tamarins*, Behaviour, 137(5).

2001: Colin Peter Groves, English-born Australian biological anthropologist, comprehensively reconstructed *primate classification* from the species level upward, such as published in his work *Primate Taxonomy*, Washington D.C.: Smithsonian Institution Press. \ László Kordos, Hungarian geologist, and David Begun, Canadian anthropologist, published their works *Primates from Rudabánya: allocation of specimens to individuals, sex and age categories*, and *A new cranium of Dryopithecus from Rudabánya, Hungary*, Journal of Human Evolution, 40 (1) and 41 (6) respectively.

2002: Kennet Christopher Beard, US palaeontologist, included the fossil *Parapithecus* in his 'Basal anthropoids', as published by Harwig Walter ed., *The Primate Fossil Record*, Cambridge University Press.

2005: Elwyn Seymour Simons, Tab Rasmussen and Daniel Gebo, US primate anthropologists, published their article *A new species of Propliopithecus from Fayum, Egypt*, American Journal of Physical Anthropology, 73(2).

2006: Francine Neago, French primatologist and conservationist specialized in orang-utans, became best known for her orangutan language televised programs and preservation of endangered great apes; she wrote the book *Orphaned orangutan raised by humans*, and developed a computer system to teach language to orang-utans, as well as organized and directed the world's first *orangutan language study*.

2009: Mark Hauser, Katherine McAuliffe and Peter Blake, US evolutionary biologists, based on the previous observation that cotton-top tamarins can recognize themselves in a mirror, these biologists continued the research and published their results as *Evolving the ingredients for reciprocity and spite*, Philosophical Transactions of the Royal Society B: Biological Sciences, 364.

2009-10: Iyad Zalmout, William Sanders, Laura MacLatchy, Gregg Gunnell, Yahya Al-Mufarreh, Mohammad Ali, Abdul-Azziz Nasser, Abdu Al Masari, Salih Al-Sobhi, Ayman Nadhra, Adel Matari, Jeffrey Wilson and Philip Gingerich, Saudi Arabian and US palaeontologists and anthropologists, first I Zalmout discovered fossil remains from *Saadanius hijazensis* near Mecca in western Saudi Arabia, and then together with the other ones analyzed these remains, and published their work *New Oligocene primate from Saudi Arabia and the divergence of apes and Old World monkeys*, Nature, 466.

2009-13: Brandon Keim, US researcher on animal communication, produced a work entitled *Rudiments of Language Discovered in Monkeys*, Wiredscience, and published later findings in the Proceedings of the National Academy of Sciences, describing *monkeys' syntax*, or principles of word sequence and sentence structure.

2010: Muhammad Spocter, William Hopkins, Amy Garrison, Amy Bauernfeind, Cheryl Stimpson, Patrick Hof and Chet Sherwood, US anthropologists, psychologists and primatologists, published their work *Wernicke's area homologue in chimpanzees (Pan troglodytes) and its relation to the appearance of modern human language*, Proceedings of the Royal Society, Biological sciences.

2011: Robin Dunbar, British psychologist, studied primates and their behaviours, and wrote the book *Grooming, Gossip and the Evolution of Language*, Kindle Edition, Faber and Faber.

3. Hominids and roots of speech

3.1. Differentiation of hominids

The primate order comprises two suborders: **STREPSIRRHINI** (primates with moist noses), and **HAPLORRHINI** (primates with dry noses).
The Haplorrhini suborder is composed of two infraorders: **TARSIIFORMES** (tarsiers), and **SIMIIFORMES** (anthropoids).
Simiiformes infraorder comprises two parvorders: **_Platyrrhini_** (New World monkeys), and **_Catarrhini_** (Old World monkeys, apes and humans).
Catarrhini parvorder is composed of two superfamilies: **Cercopithecoidea** (Old World monkeys), and **Hominoidea** (apes and humans).
Hominoidea superfamily comprises two families: **_Hylobatidae_** (lesser apes), and **_Hominidae_** (hominids: great apes and humans).
Hominidae family consists of three subfamilies: _Ponginae_ (orang-utans), _Gorillinae_ (gorillas), and _Homininae_ (chimpanzees, bonobos and humans).
Homininae subfamily comprises: _chimpanzees and bonobos_, and _humans_.

On the basis of Primate order taxonomy, studies concerning the human ancestry became a fascinating field of interest and activity, especially since the origin of the theory of evolution by natural selection (CR Darwin, 1971-73), and had an accelerated course in the last fifty years (V Sarich and A Wilson, 1967; REF Leakey, 1981-94; R Merkle, 1989; BH Smith, 1991; I Tattersall, 1993-98; R McKie and C Stringer, 1996; V Morell, 1996-2013; W Bodmer, R McKie and D Drubach, 2000; J Carey, 2003; M Tomasello and E Herrmann, 2008-10; HM McHenry, 2009).
Today, the evolution of hominids is studied by disciplines including archaeology, primatology, palaeoanthropology, physical anthropology, ethology, linguistics, evolutionary psychology, embryology, genetics and neurology.

Hominids are differentiated from apes by characteristics such as: _(i)_ apes live in the trees and also on the land, while hominids are exclusively terrestrial; _(ii)_ apes are not fully bipedal like hominids; _(iii)_ a hominid has a pair bonding with another member and also a home base, while an ape lacks this developmental trait; _(iv)_ apes do not cope with food sharing, while hominids have been shown to care and share food with other members of the group; _(v)_ hominids, _e.g. Homo habilis_, were discovered with the first making and use of tools and also fire, but apes have not found the ability to make and use tools; *(vi)* apes have a relatively smaller brain than hominids, whence other important differences between the two groups.

The bones of more than 500 hominins have been found, from which the evolution of their related species was better understood. There were considerable anatomical differences among the early hominins, but meanwhile they shared some important traits. From about 3 million years ago most of them were nearly as efficient at bipedal locomotion as humans. Unlike apes, the bones of their pelvis were shortened from top to bottom and bowl-shaped, making the pelvis more stable for weight support when standing upright or moving bipedally. By difference, the longer ape pelvis is adapted for quadrupedal locomotion. Early hominin leg and foot bones were also more similar to ours than to those of apes.

Bipedal locomotion may have been an adaptation to living in a mixed woodland and grassland environment, where bipedalism was useful to see over long distances when moving in areas covered with tall grasses. The upright posture also potentially enables to dissipate excess body heat and reduce the absorption of heat from the sun because less skin has a direct exposure to ultraviolet radiation during the hottest hours of the day.

Bipedal animals usually can walk greater distances because less energy is expended with their longer strides, enabling them to look for food throughout vast areas.

3.2. Main characteristics

On the lineage from early hominids to modern humans, there were significant adaptations and changes, such as:
- Bipedalism (use of two feet for walking),
- Increased brain size (encephalization),
- Lengthened ontogeny (gestation and infancy),
- Decreased sexual dimorphism (less phenotypic difference between males and females),
- Evolution of a power and precision grip (starting with Homo erectus),
- Increased capability of vision rather than smell,
- Smaller gut (alimentary canal),
- Loss of body hair,
- Development of sweat glands,
- Modification of dental arcade from u-shaped to parabolic,
- Development of chin (at Homo sapiens only),
- Lowering position of larynx.

The earliest bipedal hominin is considered to be either Sahelanthropus or Orrorin, followed by the fully bipedal Ardipithecus, then by Australopithecus, and later by the genus Homo.

Hominid communication and speech gradually developed from signalling and instinctive (emotional) vocalization through an intentional (controlled) vocalization to simple and then complex forms of speech.

On the human lineage, genetic differences took place when new forms appeared, in million years ago, as: *Hominidae*, 9 - 7; *Homo*, 6 - 4; and *Homo sapiens*, about 0.2. The X chromosomes of humans and chimpanzees appear to have diverged about 1.2 million years later than the other chromosomes. Humans have 23 pairs of chromosomes, while chimpanzees, gorillas, and orang-utans have 24. This difference is because the human *chromosome 2* is a fusion of two *chromosomes, 2a and 2b*, that remained separate in the other primates.

Local populations of Homo neanderthalensis and Homo floresiensis became extinct because either changing environmental conditions, or interbreeding with Homo sapiens, so that nearly all modern non-African humans have between 1% and 4% of their DNA derived from Neanderthals' or Floresiens' DNA, and up to 6% of their genome common with Denisova hominin.

Human evolutionary success derived from a relatively larger brain with well-developed neocortex, prefrontal cortex and temporal lobes, as well as from modifications of axons in the neocortex arising from extrinsic neurotransmitter systems, and molecular adaptations for high neuronal activity level. The increasing complexity and efficiency of *Homo* brain led to development of abstract reasoning, language, capacity of problem solving, sociality and culture through social learning.

An attempt to explain hominid evolution was mainly based on the *encephalization*, as the amount of brain mass in relation to an animal's total body mass, such as

$$b_w = C \cdot B_w{}^r,$$

where b_w is the brain weight, C is the cephalisation factor, B_w is the body weight, and r is the exponential constant considered 0.28 for primates and either 0.56 or 0.66 for mammals in general. This quotient has values of 2.09 for Rhesus monkey, 2.49 for chimpanzee, 5.31 for dolphin, and 7.44 for human.

Human species developed a much larger brain than that of other primates - typically 1330 cm^3 in modern humans, over twice the size of that of a chimpanzee or gorilla. The increase in volume over time has affected areas within the brain unequally - the *temporal lobes*, which contain centres for language processing, have increased disproportionately, as has the *prefrontal cortex* which has been related to complex decision-making and moderating social behaviour. The encephalization has been tied to an increasing emphasis on meat in the diet, or with the development of cooking, and it has been proposed that intelligence increased as a response to an increased necessity for solving social problems as human society became more complex.

The time when two species or other taxa diverged, i.e. the time of occurrence of events called speciation or radiation, can be estimated using the *molecular clock* technique (É Zuckerkandl and L Pauling, 1962), especially applied to nucleotide sequences for DNA or amino acid sequences for proteins, on the basis of fossil constraints and rates of molecular change. This technique, also called *gene clock* or *evolutionary clock*, was improved (D San Mauro and A Agorreta, 2010), enabling to calculate the times of divergence during the great ape-hominid-human molecular evolution, such as: *(i)* great apes separated from gibbons about 16 - 15 million years ago; *(ii)* Homininae diverged from orang-utans c.13 million years ago; *(iii)* bonobos and chimpanzees split from gorillas c.10 million years ago; *(iv)* Sahelanthropus speciated from chimpanzees c.7 million years ago; and further the human lineage continued.

Usually, the hominids are characterized by their brain volume (cranial capacity) q expressed in cubic centimetres, with increasing average values such as:

Hominids	Brain volume q (cm^3)
Sahelanthropus tchadensis	280 - 320
Paranthropus troglodytes	400 - 410
Paranthropus aethiopicus	420 - 440
Australopithecus afarensis	438 - 450
Australopithecus garhi	440 - 460
Australopithecus africanus	452 - 600
Paranthropus boisei	500 - 530
Paranthropus robustus	520 -540
Homo habilis	550 - 690
Homo georgicus	600 - 730
Homo rudolfensis	700 - 789
Homo ergaster	800 - 871
Homo erectus	870 - 1075
Homo heidelbergensis	1100 - 1300
Homo neanderthalensis	1200 - 1800

Among the primates, the hominids have relatively higher values of encephalization quotient (*EQ*) such as: *Australopithcus afarensis* 2.5, *Paranthropus boisei* 2.7, *Paranthropus robustus* 3.0, *Homo ergaster* 3.3, *Homo habilis* 3.6, *Homo erectus* 3.61, *Homo heidelbergensis* 5.26, *Homo neanderthalensis* 5.5, and, for comparison, *Homo sapiens* 5.8 (P Young, 2006; M Power and J Schulkin, 2009).

Human lineage has unfolding on a period of 9 - 7 million years, when the average brain volume increased from $280...320 \approx 300$ cm^3 = $3 \cdot 10^{-4}$ cubic metres to $1200...1800 \approx 1500$ cm^3 = $1.5 \cdot 10^{-3}$ cubic metres; and lifespan from about 25 years $\approx 7.89 \cdot 10^8$ seconds to $65...69 \approx 67$ years $\approx 2.11 \cdot 10^9$ seconds (BH Smith, 1991; J Carey, 2003). During the same interval of time, the brain power also increased up to $20...25 \approx 22.5$ watts (R Merkle, 1989; D Drubach, 2000). As in the latest stage of human lineage the average brain volume of $1.5 \cdot 10^{-3}$ cubic metres corresponds to the average brain power of 22.5 watts, it follows that in its earliest stage the average brain volume of $3 \cdot 10^{-4}$ cubic metres corresponds to an average brain power of ($3 \cdot 10^{-4}$ cubic metres)$\cdot$(22.5 watts)/($1.5 \cdot 10^{-3}$ cubic metres) ≈ 4.5 watts. Furthermore, the average brain energy, calculated as the product of average brain power and the average lifespan, increased during the process from (4.5 watts)$\cdot$($7.89 \cdot 10^8$ seconds) $\approx 3.55 \cdot 10^9$ joules to (22.5 watts)$\cdot$($2.11 \cdot 10^9$ seconds) $\approx 4.748 \cdot 10^{10}$ joules; and meanwhile the average density of brain energy, calculated as the average density of brain energy to average brain volume, increased from ($3.55 \cdot 10^9$ joules)/($3 \cdot 10^{-4}$ cubic metres) $\approx$ 1.183·10^{13} joules per cubic metre to (4.748·10^{10} joules)/(1.5·10^{-3} cubic metres) $\approx$ 3.165·10^{13} joules per cubic metre. Using these physical parameters, a model for development of brain during the lineage of human ancestors can be completed such as:

Years ago	Average brain volume q [m^3]	Average brain power P [W]	Average lifespan λ [10^9 s]	Average brain energy $P \cdot \lambda = E$ [10^{10} J]	Average density of brain energy $E/q = \rho$ [10^{13} J/m^3]
235000	0.00150	22.50	2.110	4.748	3.165
626000	0.00120	18.00	1.911	3.440	2.867
1019000	0.00108	16.20	1.732	2.806	2.598
1412000	0.00097	14.55	1.572	2.287	2.358
1806000	0.00087	13.05	1.430	1.866	2.145
2201000	0.00079	11.85	1.305	1.546	1.957
2596000	0.00072	10.80	1.196	1.292	1.794
2993000	0.00066	9.90	1.100	1.089	1.650
3390000	0.00060	9.00	1.020	0.918	1.530
3787000	0.00055	8.25	0.953	0.786	1.429
4186000	0.00050	7.50	0.898	0.674	1.348
4585000	0.00046	6.80	0.850	0.578	1.257
4985000	0.00042	6.20	0.840	0.521	1.240
5386000	0.00039	5.85	0.815	0.477	1.223
5788000	0.00036	5.40	0.807	0.436	1.211
6190000	0.00034	5.10	0.801	0.409	1.203
6593000	0.00032	4.80	0.795	0.382	1.194
6997000	0.00031	4.65	0.792	0.368	1.187
7401000	0.00030	4.50	0.789	0.355	1.183

Many non-human primates, such as Australopithecines, have *hypoglossal canals* equal in size to those of the modern humans, but they are essentially different in structure and function, so that the size of hypoglossal canal does not imply vocal capabilities of speech or language (D DeGusta *et al.*, 1999). However, according to recent findings and studies, the hypoglossal canals of later hominids, such as Homo heidelbergensis and especially Homo neanderthalensis, are as similar to those of humans as they could have speech capabilities although less than humans.

3.3. Timeline grading

The evolution of bigger-brained, highly intelligent, and longer-living primates and hominids has been explained by the increased durations of dependency facilitating an extended period of learning (D Falk, 2004). Prelinguistic evolution in early hominins was associated with the role of bipedalism and loss of infant clinging, as well as use of vocalizations to 'keep in touch'; but true syntactic speech did not evolve until after the emergence of the genus *Homo* around two million years ago. The route from gesture to speech was marked by the transition from lexical morpheme, or intonation, to grammatical morpheme.

In the later stage of hominids' evolution, the bipedalism started to play a significant part in changes to the skull, allowing for a more L-shaped vocal tract. Furthermore, the shape of tract and a larynx positioned relatively low in the neck were necessary prerequisites for many of sounds produced by humans, particularly vowels.

Earlier human ancestors, such as *Homo habilis* and *Homo erectus*, would likely have possessed less developed forms of speech, as intermediate between the ordinary communication systems and human language. *Homo ergaster* and *Homo heidelbergensis* were the first hominins to make controlled vocalizations; the second one developing more sophisticated culture and possibly an early form of symbolic speech.

After them, *Homo neanderthalensis* possessed a hyloid bone enabling sounds similar to modern humans, and also speech abilities. Nevertheless, the _speech_ is one among a number of different methods of encoding and transmitting linguistic information, and therefore distinct from _language_ that is not necessarily spoken but might alternatively be written or signed.

The series of identified hominids on the human ancestry was preceded by *Ouranopithecus turkae* and *Ankarapithecus meteai*; started with *Sahelanthropus tchadensis* and *Orrorin tugenensis*; continued with species such as *Homo habilis* and *Homo erectus*; and ascended by *Homo heidelbergensis* and *Homo neanderthalensis* up to modern *Homo sapiens*.

According to recent discoveries and studies, other members were identified and included, so that the series of hominids can be now presented, in million years from their appearance, as follows:

82

8.7 - 7.4: *Ouranopithecus turkae*, the youngest known orangutan-like Eurasian great ape, that lived in central Anatolia, Turkey (FLP Gevaise, 1872; ES Güleç *et al.*, 2007);

7.8 - 7.5: *Ankarapithecus meteai*, a hominid skeleton found in central Anatolia, Turkey (B Alpagut *et al.*, 1996; D Begun and ES Güleç, 1998);

7.2 - 6.8: *Sahelanthropus tchadensis*, a hominin skull from a species close to chimpanzee-human divergence, found in Chad (A Beauvilain, 2003; M Brunet, 2000-01, 2006; A Klages, 2008);

7.2 - 6.9: *Sivapithecus sivalensis*, fossil remains found in the Siwalik Hills, Indian subcontinent (J Fleagle, 1988);

6.5 - 5.2: *Rudapithecus hungaricus*, fossil remains from a hominin found at Rudabánya, North Hungary (H Gábor, 1965);

6.1 - 5.7: *Orrorin tugenensis*, one of the first hominids found in Lukeino Formation, Kenya (B Senut and M Pickford, 2000-01);

5.8 - 5.2: *Ardipithecus kadabba*, an early bipedal hominid that lived in Middle Awash Valley, Ethiopia (T White *et al.*, 1997-2004);

4.4: *Ardipithecus ramidus*, one of the oldest human ancestors whose remains were found in Afar Depression, Eastern Africa (T White, 1992-95);

4.2 - 3.9: *Ardipithecus anamensis*, a hominid found at Kanapoi - Allia Bay in Kenya (M Leakey and A Walker, 1994-98);

4.0 - 3.0: *Australopithecus afarensis*, a fossil female hominid, known as 'Lucy', discovered in Afar Triangle region of Hadar, West Ethiopia (DC Johanson *et al.*, 1974-81);

3.6 - 3.0: *Australopithecus bahrelghazali*, a fossil hominin, initially called 'Abel' that lived in Chad (M Brunet *et al.*, 1995);

3.5 - 3.2: *Kenyanthropus platyops*, one of the hominins discovered in Kenya (J Erus, in M Leakey's team, 1999-2000);

3.0 -2.0: *Australopithecus africanus*, first identified by studying one of its ape-like infant part-skull found in Botswana (RA Dart, 1925);

2.7 - 2.4: *Paranthropus aethiopicus*, or *Australopithecus aethiopicus*, a fossil hominid first discovered in Ethiopia-Kenya area (B Wood and D Strait, 2004);

2.6 - 1.4: *Paranthropus boisei*, a species of *Paranthropus* that lived in Tanzania (B Wood and D Strait, 2004);

2.5: *Australopithecus garhi*, an australopithecine species whose fossils were discovered in Ethiopia (B Asfaw *et al.*, 1996-99);

2.4 - 1.5: *Homo habilis*, a species of hominin that lived in Africa, probably capable of rudimentary speech because it possessed the bulge of cerebral Broca's area (LSB Leakey *et al.*, 1964);

2.3 - 1.2: *Paranthropus robustus*, or *Australopithecus robustus*, one of the hominid fossils discovered in South Africa (R Broom, 1938-46);

2.0 - 0.4: *Homo erectus*, first known as *Pithecantropus erectus* 'Java man' that lived in Indonesian island of Java (MEFT Dubois, 1891-93); and then as *Sinanthropus pekinensis* 'Peking man', now *Homo erectus pekinensis*, found at Zhoukoudian (Chou K'ou-tien) near Beijing, China (D Black, 1827-29; F Weidenreich, 1943);

1.98: *Australopithecus sediba*, an Australopithecus identified by skeletons from the Malapa Fossil Site at Cradle of Humankind World Heritage Site in South Africa (L Berger, *et al.*, 2009-10);

1.9 - 1.89: *Homo rudolfensis*, or *Australopithecus rudolfensis*, an extinct hominin known from fossils discovered at Koobi Fora on the east side of Lake Rudolf (now Lake Turkana) in Kenya (B Ngeneo and VP Alekseyev, 1972-86);

1.8 - 1.3: *Homo ergaster*, or *African Homo erectus*, an extinct chronospecies of Homo that lived in eastern and southern Africa (CP Groves and V Mazák, 1975);

1.77 - 1.1: *Homo georgicus*, Dmansi, Georgia identified by its so-called 'Skull 5' (D Lordkipanidze, 1991-2007; C Zollikofer *et al.*, 2013);

1.2 - 0.8: *Homo antecessor*, one of the earliest known human species in Europe that lived at Atapuerca in Spain (JM Bermúdez de Castro and E Carbonell, 1997);

0.7 - 0.4: *Homo heidelbergensis*, an extinct species of genus Homo whose fossils were found in Germany, France, Greece, Italy and other European countries (O Schötensack, 1908);

0.45 - 0.35: *Homo cepranensis*, a human specimen known by its remains of skull unearthed in the Frosinone province, c.90 kilometres south-east of Rome (I Biddittu and F Mallegni, 1994-2003);

0.3 - 0.12: *Homo rhodesiensis*, a Homo species with fossil remains discovered in South, East and North Africa (AS Woodward, 1921);

0.3 - 0.025: *Homo neanderthalensis*, an extinct Homo species that lived in Eurasia, from western Europe to central and northern Asia (P-C Schmerling, 1829-31; JK Fuhlrott, 1856; W King, 1863-64);

0.2: *Homo sapiens*, the present and only surviving human species of genus *Homo* (C Linnaeus, 1758).

On the evolutionary scale of hominids, the study of their remaining fossils is primarily focused on the specific *morphologies*, such as dental morphology, vertebral column morphology, and thorax morphology (P O'Higgins, 2000). In the behavioural patterns of the early hominids, one of the most important and intriguing questions is about their *diet*. Around 4 million years ago, *Australopithecus anamensis* had thicker molar enamel, suggesting that it was able to withstand the functional demands of hard and perhaps abrasive objects in its diet. About 3 million years ago, *Australopithcus africanus* had yet another increase in postcanine tooth size, indicating that its food included quite big and abrasive items, and emphasizing its omnivory, as evidenced by isotopic studies. The presence of primitive stone tools in the fossil record shows that about 2.5 million years ago *Australopithecus garhi* was using stone implements to cut flesh off the bones of hunted animals.

Based on the series of identified hominid predecessors and members with their specific adaptations, a preliminary hominids' timeline can be displayed in sequences delimited by transitional times, in million years ago, such as:

Predecessors
\ 8.2 \ *Ouranopithecus turkae*
\ 7.8 \ *Ankarapithecus meteai*

Members
\ 7.4 \ *Sahelanthropus tchadensis*
\ 7.0 \ *Sivapithecus sivalensis*
\ 6.6 \ *Rudapithecus hungaricus*
\ 6.2 \ *Orrorin tugenensis*
\ 5.8 \ *Ardipithecus kadabba*
\ 5.4 \ ...... \ 5.0 \......
\ 4.6 \ *Ardipithecus ramidus*
\ 4.2 \ *Ardipithecus anamensis*
\ 3.8 \ *Australopithecus afarensis*
\ 3.4 \ *Kenyanthropus platyops*
\ 3.0 \ *Australopithecus africanus*
\ 2.6 \ *Homo habilis*
\ 2.2 \ *Homo erectus*
\ 1.8 \ *Homo ergaster*
\ 1.4 \ *Homo antecessor*
\ 1.0 \ *Homo heidelbergensis*
\ 0.6 \ *Homo neanderthalensis*
\ 0.2 \ *Homo sapiens* appeared.

The transitional times above, in million years ago, are converted into times from the life origin c.4200 million years ago, becoming:

4200 - 8.2 = 4191.8 \ 4200 - 7.8 = 4192.2 \

4200 - 7.4 = 4192.6 \ 4200 - 7.0 = 4193.0 \ 4200 - 6.6 = 4193.4 \

4200 - 6.2 = 4193.8 \ 4200 - 5.8 = 4194.2 \ 4200 - 5.4 = 4194.6 \

4200 - 5.0 = 4195.0 \ 4200 - 4.6 = 4195.4 \ 4200 - 4.2 = 4195.8 \

4200 - 3.8 = 4196.2 \ 4200 - 3.4 = 4196.6 \ 4200 - 3.0 = 4197.0 \

4200 - 2.6 = 4197.4 \ 4200 - 2.2 = 4197.8 \ 4200 - 1.8 = 4198.2 \

4200 - 1.4 = 4198.6 \ 4200 - 1.0 = 4199.0 \ 4200 - 0.6 = 4199.4 \

4200 - 0.2 = 4199.8 \ the present time.

In the primates' timeline (see table at the end of subchapter 2.3) the latest $N/10^2 = \beta$ sequences are delimited by values of argument \1.86\ ...\1.87\...\1.88\...\1.89\ corresponding to the evolution of hominids, which can be detailed by $N/10^3 = \gamma$ sequences in the succession of identified hominids' predecessors and members. As the final time had been estimated as $t_\bullet = 4400$ million years from the origin of life, the sequences marking the evolution of hominids are delimited by transitional times $t_\gamma = t_\bullet \cdot tanh(\gamma)$ calculated, in million years from the same origin, as follows:

$$t_{1.860} = t_\bullet \cdot tanh(1.860) \approx (4400) \cdot (0.9526788) \approx 4191.787;$$
$$t_{1.861} = t_\bullet \cdot tanh(1.861) \approx (4400) \cdot (0.9527712) \approx 4192.193;$$
$$t_{1.862} = t_\bullet \cdot tanh(1.862) \approx (4400) \cdot (0.9528633) \approx 4192.599;$$
$$t_{1.863} = t_\bullet \cdot tanh(1.863) \approx (4400) \cdot (0.9529553) \approx 4193.003;$$
$$t_{1.864} = t_\bullet \cdot tanh(1.864) \approx (4400) \cdot (0.9530471) \approx 4193.407;$$
$$t_{1.865} = t_\bullet \cdot tanh(1.865) \approx (4400) \cdot (0.9531387) \approx 4193.810;$$
$$t_{1.866} = t_\bullet \cdot tanh(1.866) \approx (4400) \cdot (0.9532301) \approx 4194.212;$$
$$t_{1.867} = t_\bullet \cdot tanh(1.867) \approx (4400) \cdot (0.9533214) \approx 4194.614;$$
$$t_{1.868} = t_\bullet \cdot tanh(1.868) \approx (4400) \cdot (0.9534125) \approx 4195.015;$$
$$t_{1.869} = t_\bullet \cdot tanh(1.869) \approx (4400) \cdot (0.9535034) \approx 4195.415;$$
$$t_{1.870} = t_\bullet \cdot tanh(1.870) \approx (4400) \cdot (0.9535941) \approx 4195.814;$$
$$t_{1.871} = t_\bullet \cdot tanh(1.871) \approx (4400) \cdot (0.9536847) \approx 4196.213;$$
$$t_{1.872} = t_\bullet \cdot tanh(1.872) \approx (4400) \cdot (0.9537751) \approx 4196.610;$$
$$t_{1.873} = t_\bullet \cdot tanh(1.873) \approx (4400) \cdot (0.9538653) \approx 4197.007;$$
$$t_{1.874} = t_\bullet \cdot tanh(1.874) \approx (4400) \cdot (0.9539554) \approx 4197.404;$$
$$t_{1.875} = t_\bullet \cdot tanh(1.875) \approx (4400) \cdot (0.9540453) \approx 4197.799;$$
$$t_{1.876} = t_\bullet \cdot tanh(1.876) \approx (4400) \cdot (0.9541350) \approx 4198.194;$$
$$t_{1.877} = t_\bullet \cdot tanh(1.877) \approx (4400) \cdot (0.9542245) \approx 4198.588;$$
$$t_{1.878} = t_\bullet \cdot tanh(1.878) \approx (4400) \cdot (0.9543139) \approx 4198.981;$$
$$t_{1.879} = t_\bullet \cdot tanh(1.879) \approx (4400) \cdot (0.9544031) \approx 4199.374;$$
$$t_{1.880} = t_\bullet \cdot tanh(1.880) \approx (4400) \cdot (0.9544921) \approx 4199.765;$$
$$t_{1.881} = t_\bullet \cdot tanh(1.881) \approx (4400) \cdot (0.9545810) \approx 4200.156.$$

Therefore, the hominids' sequences are integrated in the general evolution of life, as shown in the table below.

$N/10^3 = \gamma$	Time (million years)		Sequences in evolution of hominids
	from origin $t_y=t_\bullet \cdot tanh(\gamma)$	from present t_y - 4200	
	4400	+200	
...	...	...	
1.881	4200.156	+0.156	_
1.880	4199.765	-0.235	_Homo sapiens
1.879	4199.374	-0.626	_Homo neanderthalensis
1.878	4198.981	-1.019	_Homo heidelbergensis
1.877	4198.588	-1.412	_Homo antecessor
1.876	4198.194	-1.806	_Homo ergaster
1.875	4197.799	-2.201	_Homo erectus
1.874	4197.404	-2.596	_Homo habilis
1.873	4197.007	-2.993	_Australopithecus africanus
1.872	4196.610	-3.390	_Kenyanthropus platyops
1.871	4196.213	-3.787	_Australopithecus afarensis
1.870	4195.814	-4.186	_Ardipithecus anamensis
1.869	4195.415	-4.585	_Ardipithecus ramidus
1.868	4195.015	-4.985	_Ardipithecus ramidus
1.867	4194.614	-5.386	_... ...
1.866	4194.212	-5.788	_... ...
1.865	4193.810	-6.190	_Ardipithecus kadabba
1.864	4193.407	-6.593	_Orrorin tugenensis
1.863	4193.003	-6.997	_Rudapithecus hungaricus
1.862	4192.599	-7.401	_Sivapithecus sivalensis
1.861	4192.193	-7.807	_Sahelanthropus tchadensis
1.860	4191.787	-8.213	_Ankarapithecus meteai
...	...	...	_Ouranopithecus turkae
0	0	-4200	_Ouranopithecus turkae

References

1758: <u>Carolus Linnaeus</u>, Swedish botanist, zoologist and physician, introduced the *Linnaean taxonomy*, now known as *binomial nomenclature*, gave the name *Homo* (from Latin *homo* 'human being') to the biological genus to which humans belong and coined the species binomial *Homo sapiens* (meaning in Latin 'wise man'); his scientific work was crowned by the 10th edition of *Systema Naturae* published as *Systema Naturae Per Regna Tria Naturæ, Secundum Classes, Ordines, Genera, Species Cum Characteribus, Differentiis, Synonymis, Locis* 'System of nature through the three kingdoms of nature, according to classes, orders, genera and species, with characters, differences, synonyms, places', Tomus I, Decima, Reformata.

1829-31: <u>Philippe-Charles Schmerling</u>, Dutch-Belgian prehistorian and palaeontologist, discovered the first *Neanderthal skull*, and published *Cavernes à ossements fossiles, découvertes jusqu' à ce jour dans la province de Liège* 'Caves with fossil bones, discovered until today in the province of Liège', in Philippe Vandermaelen, 'Dictionnaire géographique de la province de Liège', Bruxelles, Éstablissement géographique.

1856: <u>Johann Karl Fuhlrott</u>, German mathematician and anthropologist, first recognized the fossil called *Neanderthal man*, discovered at Neanderthal in North Rhine-Westphalia, Germany.

1863-64: <u>William King</u>, Anglo-Irish geologist, first proposed that bones found at Neanderthal in Germany were not of human origin, but of a distinct species called *Homo neanderthalensis*, and published his article *The reputed fossil man of the Neanderthal*, Quarterly Journal of Science, 1.

1871-73: <u>Charles Robert Darwin</u>, English naturalist and originator of the theory of evolution by natural selection, wrote and published treatises including *The Descent of Man and Selection in relation to Sex*, and *The Expression of the Emotions in Man and Animals*.

1891-93: <u>Marie Eugène François Thomas Dubois</u>, Dutch paleoanthropologist and geologist, worked in Indonesia and explored some of its islands, especially Java where he discovered a hominid fossil, called by him 'Java man', or *Pithecanthropus erectus*, and later redesigned as one of the first known specimens of *Homo erectus*.

1908: <u>Otto Schötensack</u>, German industrialist and anthropologist, described a jaw of *Homo heidelbergensis* that was found near Heidelberg in Germany, and published his work entitled *Der Unterkiefer des Homo heidelbergensis aus den Sanden von Mauer bei Heidelberg* 'The lower jaw of the Homo heidelbergensis out of the sands of Mauer near Heidelberg', Leipzig, Wilhelm Engelmann.

1921: <u>Arthur Smith Woodward</u>, English palaeontologist, first described the fossil *Homo rhodesiensis*, making it public by his writing *A New Cave Man from Rhodesia, South Africa*, Nature, 108.

1925: <u>Raymond Arthur Dart</u>, South African anatomist, worked on anthropology and described an ape-like infant part-skull found in Botswana which he considered to be a human ancestor; his findings were published as *Australopithecus africanus, the Man-Ape of South Africa*, Nature, 115.

1827-29: <u>Davidson Black</u>, Canadian paleoanthropologist, working as director of the Anatomy Department of Beijing Union Medical College, he analyzed some teeth and then a lower jaw and skull fragments from a 'Peking man' just discovered at Zhoukoudian (Chou K'ou-tien) near Beijing, China, and named that man *Sinanthropus pekinensis*, now called *Homo erectus pekinensis*; the results of his researches were published as *Preliminary Notice on the Discovery of an Adult Sinanthropus Skull at Chou Kou Tien*, Bulletin of Geological Society of China, VIII (3).

1938-46: <u>Robert Broom</u>, Scottish South African doctor and palaeontologist, discovered and analyzed fossil remains from *Paranthropus robustus* and then published the book *The South African fossil ape-man: the Australopithecinae*, Transvaal Museum in Pretoria.

1943: <u>Franz Weidenreich</u>, Jewish German anatomist and physical anthropologist, continued the studies of D Black concerning the fossil remains from *Sinanthropus pekinensis* 'Peking man', and published his book *The Skull of Sinanthropus pekinensis: A Comparative Study on a Primitive Hominid Skull*, The National Geological Survey of China.

1962: <u>Émile Zuckerkandl</u>, Austrian-born French biologist, and <u>Linus Pauling</u>, US biochemist, noticed that the number of *amino acid differences in haemoglobin* between different lineages changes are roughly linear with time, as estimated from fossil evidence; and published the results of their research as an article entitled *Molecular disease, evolution, and genic heterogeneity*, in M Kasha and B Pullman eds., 'Horizons in Biochemistry', Academic Press, New York.

1964: <u>Louis Seymour Bazet Leakey</u>, British palaeontologist and anthropologist, <u>Phillip Valentine Tobias</u>, South African paleoanthropologist, and <u>John Russell Napier</u>, British primatologist and paleoanthropologist, studied the fossil remains from *Homo habilis*, and wrote the article *A new species of the genus Homo from Olduvai Gorge*, Nature, 4.

1965: <u>Hernyák Gábor</u>, Hungarian mining geologist, researching the sedimentary formation in the former mining area at Rudabánya, Northern Hungary, he discovered fossil remains from the later called specimen *Rudapithecus hungaricus*.

1967: <u>Vincent Sarich</u>, US anthropologist, and <u>Allan Wilson</u>, US biochemist, measured the strength of immunological *cross-reactions of blood serum albumin* between pairs of creatures, including humans and African apes (chimpanzees and gorillas), and estimated the divergence time of humans and apes up to 5 million years ago; they published *Immunological time scale for hominid evolution*, Science, 158.

1972-86: <u>Bernard Ngeneo</u>, member of the team led by REF Leakey and M Leakey, and <u>Valeri Pavlovich Alekseyev</u>, Russian anthropologist, the first of them discovered a fossil of *Australopithecus rudolfensis* on the east side of Lake Rudolf (now Lake Turkana), and the last one proposed the binomial name *Homo rudolfensis* in his published work *The Origin of the Human Race*, Progress Publishers.

1974-81: <u>Donald Carl Johanson</u>, <u>Maurice Taieb</u> and <u>Yves Coppens</u>, US palaeontologists, discovered and studied the 3.2 million-year-old fossil of a female hominid *Australopithecus afarensis*, known as 'Lucy' that was found in the Afar Triangle region of Hadar, Ethiopia, and later popularized in the book *Lucy: The Beginning of Humankind*, New York: Simon and Schuster.

1975: <u>Colin Peter Groves</u>, Australian anthropologist, and <u>Vratislav Mazák</u>, Czech biologist, after studying fossil remains from *African Homo erectus*, they introduced the binomial name *Homo ergaster* (Greek ἐργαστήρ 'workman') in the published work *An approach to the taxonomy of the Hominidae: Gracile Villafranchian hominids of Africa*, Casopis pro Mineralogii a Geologii, 20.

1981-94: <u>Richard Erskine Frere Leakey</u>, Kenyan paleoanthropologist, wrote and published a series of books including *The Making of Mankind*, Michael Joseph, London; *Origins Reconsidered*, Little, Brown & Co., London; and *The Origin of Humankind*, Weidenfeld & Nicolson, London.

1988: <u>John Fleagle</u>, US researcher, made detailed studies on functional and evolutionary approaches of *comparative primate data*, and wrote the book entitled *Primate Adaptation and Evolution*, Academic Press: New York.

1989: <u>Ralph Merkle</u>, US computer scientist, wrote the paper *Energy Limits to the Computational Power of Human Brain*, Foresight Update, 6.

1991: <u>Bennet Holly Smith</u>, US biological anthropologist, published her work *Dental Development and the Evolution of Life History in Hominidae*, American Journal of Physical Anthropology, 86.

1991-2007: <u>David Lordkipanidze</u>, Georgian scientist, discovered the so-called 'Skull 5' of *Homo georgicus* in Dmanisi, Georgia; his discovery was followed by analyses carried out by Swiss researchers, who established that the 1.8 million year-old skull shows, unlike any other *Homo* remains on record, that it combines a long face, massive jaw and large teeth with a small braincase, just about a third the size of that found in modern humans and no longer than those of much more primitive African fossils; the results of their research were published as *Postcranial evidence from early Homo of Dmanisi, Georgia*, Nature, 449.

1992-95: <u>Tim White</u>, US anthropologist, researching for primate fossils in Africa, discovered the *Ardipithecus ramidus* in Ethiopia, dated 4.45 - 4.35 million years ago, which is considered one of the *oldest human ancestors*.

1993-98: <u>Ian Tattersall</u>, British-born US paleoanthropologist, published a series of books including *The Human Odyssey*, Prentice Hall, New York; *The Fossil Trail*, and *Becoming Human*, Oxford University Press.

1994-98: <u>Meave Leakey</u>, London-born Kenyan paleoanthropologist, and <u>Allan Walker</u>, US anthropologist, discovered *Australopithecus anamensis*, which was made known by their published works such as *New four-million-year-old hominid species from Kanapoi and Allia Bay, Kenya*, Nature 376; *Early Hominid Fossils from Africa*, Scientific American, 276 (6); and *New specimens and confirmation of an early ape for Australopithecus anamensis*, Nature, 393.

1994-2003: <u>Italo Biddittu</u> and <u>Francesco Mallegni</u>, Italian anthropologists, first of them discovered a human-like skull, initially called 'Ceprano man' after the nearby town situated c. 90 kilometres south-east of Rome; and the second one analyzed the fossil remains and published *Homo cepranensis sp. nov. and the evolution of African-European Middle Pleistocene hominids*, Comptes Rendus Palevol. 2(2).

1995: <u>Michel Brunet</u>, <u>Alain Beauvilan</u>, <u>Yves Coppens</u>, <u>Emile Heintz</u>, <u>Aladji Moutaye</u> and <u>David Pilbeam</u>, French-American-English team of anthropologists, discovered and studied the fossil of *Australopithecus bahrelghazali*, initially called 'Abel', in the Bahr el Ghazal valley near Koro Toro, in Chad, which was made known by their published work *The first australopithecine 2500 kilometres west of the Rift Valley (Chad)*, Nature, 378.

1996: <u>Robin McKie</u>, English Science Editor of the Observer, and <u>Chris Stringer</u>, English palaeontologist, analytically examining the new discoveries in human history, they wrote the book entitled *African Exodus: The Origins of Modern Humanity*. \ <u>Berna Alpagut</u>, Turkish, <u>Peter Andrews</u>, British, <u>Mikael Fortelius</u>, Finnish, <u>John Kappelman</u>, US, <u>Ilhan Temizsoy</u>, Turkish, <u>Hürkan Çelebi</u>, Turkish, and <u>William Lindsay</u>, British, anthropologists, studied the fossil remains of *Ankarapithecus meteai*, and published their results in the article entitled *A new specimen of Ankarapithecus meteai from the Sinap Formation of central Anatolia*, Nature, 382.

1996-99: <u>Berhane Asfaw</u>, Ethiopian, <u>Tim White</u>, US, <u>Owen Lovejoy</u>, US, <u>Bruce Latimer</u>, US, <u>Scott Simpson</u>, US, and <u>Gen Suwa</u>, Japanese, palaeontologists and anthropologists, discovered and studied the fossil *Australopithecus garhi*, then published the article *Australopithecus garhi: a new species of early hominid from Ethiopia*, Science, 284.

1996-2013: <u>Virginia Morell</u>, US natural historian and science-writer, published a series of books including *Ancestral Passions: The Leakey Family and the Quest for Humankind's Beginnings*, Touchstone, New York; *Animal Wise: The Thoughts and Emotions of Our Fellow Creatures*, Kindle Edition; and *Blue Nile: Ethiopia's River of Magic and Mystery*, Adventure Press.

1997: <u>José María Bermúdez de Castro</u> and <u>Eudald Carbonell</u>, Spanish anthropologists, took part in discovery of fossil *Homo antecessor*, one of the earliest known human species in Europe, found at the site Sima del Elefante, Atapuerca, Spain; and wrote the paper *Size variation in Middle Pleistocene humans*, Science, 277.

1997-2004: <u>Tim White</u>, US anthropologist, together with <u>Gen Suwa</u> and <u>Yohannes Haile-Selassie</u>, African anthropologists, discovered and studied the first known bipedal called *Ardipithecus kadabba*, in Middle Awash Valley, Ethiopia, which was dated from 5.6 - 5.4 to 5.2 million years ago, as the earliest human ancestor.

1998: David Begun, Canadian paleoanthropologist, and Erksin Savas Güleç, Turkish anthropologist, continuing the researches on the *Ankarapithecus* found in central Anatolia, they showed the implications and links of the studied fossil within the article *Restoration of the Type and Palate of Ankarapithecus meteai: Taxonomic and Phylogenetic Implications*, American Journal of Physical Anthropology, 105.

1999: David DeGusta, Henry Gilbert and Scott Turner, US anthropologists, after studying speech capacities of a number of hominids, especially Australopithecines, they published a paper entitled *Hypoglossal canal size and hominid speech*, Proceedings of the National Academy of Sciences, 96.

1999-2000: Justus Erus, Kenyan anthropologist, followed by Meave Leakey, Kenyan, Fred Spoor, English, Hank Brown, US, Patrick Gathogo, US, Christopher Kiarie, Kenyan, Louise Leakey, Kenyan, and Ian McDougall, Australian, anthropologists, first discovered remains of a hominin in the Lake Turkana, Kenya, then they studied and named it *Kenyanthropus platyops*, and published their results as *New hominin genus from eastern Africa shows diverse middle Pliocene lineages*, Nature, 410.

2000: Walter Bodmer, English geneticist, and Robin McKie, English Science Editor of the Observer, produced *The Book of Man: The Quest To Discover Our Genetic Heritage*, Orion Audio Books. \ Robin McKie, English Science Editor of the Observer, published his book *Ape●Man: The Story of Human Evolution*, BBC Worldwide Ltd. \ Daniel Drubach, US neurologist and psychobiologist, produced a work entitled *The Brain Explained*, New Jersey: Prentice-Hall. \ Paul O'Higgins, British evolutionary anatomist, wrote an article entitled *The study of morphological variation in hominid fossil record: biology, landmarks and geometry*, Journal of Anatomy, 197(1).

2000-01: Brigitte Senut and Martin Pickford, French palaeo-anthropologists, discovered so called *Orrorin tugenensis* in Tugen Hills of Kenya, dated 6.1 - 5.7 million years ago, which may been similar to our early ancestors; and published their work *First hominid from the Miocene (Lukeino Formation, Kenya)*, Comptes Rendus de l'Académie de Sciences, 332 (2). \ Michael Brunet, French palaeontologist, excavated in Chad where he found a 7.2 million year old skull of the so-named *Sahelanthropus tchadensis*, considered by him to be a hominin, but then argued to be in the human evolutionary line, and reconsidered as a common ancestor with chimpanzees.

2003: Alain Beauvilain, French anthropological researcher, carried out field work related to *Sahelanthropus* in Chad, and published the book *Toumaï, l'aventure humaine*, La Table Ronde, Paris. \ James Carey, US biodemographer, wrote the book entitled *Longevity: The Biology and Demography of Life Span*, Princeton University Press.

2004: Bernard Wood and David Strait, US anthropologists, analyzed the former discovered fossil of genus *Paranthropus*, identified its species

Paranthropus boisei in Tanzania, and wrote *Patterns of resource use in early Homo and Paranthropus*, Journal of Human Evolution, 46(2). \ Dean Falk, US anthropologist, spent much time studying the evolution of prelinguistic behaviour, and made known the results of her work by an article entitled *Prelinguistic evolution in early hominins: Whence motherese?*, Behaviour and Brain Science, 27.

2006: Michael Brunet, French anthropologist, after researching the site and the skull of *Sahelanthropus tchadensis* in Chad, he wrote the book *D'Abel à Toumaï. Nomade, chercheur d'os*, Odile Jacob, Paris. \ Patrick Young, US creationist chemist, published his work *Body Mass Estimates and Encephalization Quotients: A Fresh Look at the Australopithecines and Homo habilis*, Creation Research Society Quarterly Journal, 42(4).

2007: Erksin Savas Güleç, Ayla Sevin, Cesur Pehlevan and Ferhat Kaya, Turkish anthropologists, unearthed *Ouranopithecus turkae*, and published their findings under the title *A new great ape from the late Miocene of Turkey*, Anthropological Science, 115(2).

2008: Arthur Klages, US anthropologist, published his paper *Sahelanthropus tchadensis: An Examination of its Hominin Affinities and Possible Phylogenetic Placement*, Totem: The University of Western Ontario Journal of Anthropology, 16(1).

2008-10: Michael Tomasello and Esther Herrmann, German evolutionary anthropologists, made comparative studies of apes and humans, and published their work *Ape and Human Cognition: What's the Difference?*, Current Directions in Psychological Science, 19.

2009: Henry Malcom McHenry, US anthropologist, researched the human ancestry and wrote the article *Human Evolution*, that was inserted in the book *Evolution: The First Four Billion Years*, Michael Ruse & Joseph Travis, eds. \ Michael Power and Jay Schulkin, US evolutionary neuroscientists, developed bio-neurological researches, and wrote the book *Evolution of Obesity*, Baltimore: John Hopkins University Press.

2009-10: Lee Berger, South African, Darryl de Ruiter, US, Steven Churchill, US, Peter Schmid, Swiss, Kristian Carlson, US, Paul Dirks, Australian, and Job Kibii, South African, paleoanthropologists, first discovered fossil remains from a juvenile male, and then a total of six skeletons belonging to *Australopithecus sediba* (in local Sotho language, *sediba* meaning 'natural spring, or well'); the results of their work were published as *Australopithecus sediba: A New Species of Homo-like Australopith from South Africa*, Science, 328.

2010: Diego San Mauro and Ainhoa Agorreta, Spanish evolutionary biologists, made researches on the precision of *molecular clock* technique, and published their work *Molecular systematics: a synthesis of the common methods and the state of knowledge*, Cellular & Molecular Biology Letters, 15(2).

2013: Christoph Zollikofer, Swiss neurobiologist, used the computer-assisted anthropology to investigate patterns of morphological variability and evolutionary diversification in fossil and extant primates, computational modelling of morphogenetic process, and development of image-based analytical tools for anthropology, applying these modern techniques in the case of *Homo georgicus* on the basis of its 'Skull 5' previously discovered; the results were published by Jeremy Kaplan as *1.8-million-year-old skull shakes mankind's family tree*, and popularized by the Live Science contributor Charles Choi who published *Oddball 'Skull 5' Fossil Suggests Early Humans Belonged to Same Species*.

4. Humans and language

4.1. Ascendancy of humans

There was a long transitional process from the hominid behaviour and communication-speech to the human characteristics and language, that sets up a central part of the *anthropology* (from Greek ἄνθρωπος 'man' and λογία 'study'), which is defined as the science of man in its widest sense; the study of human origins, societies and culture; or the study of humankind from past to present.

Anthropology is drawn and built upon knowledge from social and biological sciences, as well as humanities and the natural sciences. As one of its sub-fields, the paleoanthropology combines the disciplines of palaeontology and physical anthropology, focusing on the study of ancient humans as found in fossil hominid evidence such as petrified bones and footprints.

Biological anthropology and *physical anthropology* are synonymous terms to describe anthropological research focused on the study of humans and non-human primates in their biological, evolutionary, and demographic viewpoints. It examines the biological and social factors that have affected the evolution from hominids to humans on the basis of contemporary genetic and physiological variation.

Further, the anthropology has diversified into numerous branches, such as social, biological, archaeological, linguistic, musical, visual, economic, applied and development, medical, nutritional, psychological, cognitive, transpersonal, political, legal, digital, ecological, environmental, historical, and urban anthropologies, as well as anthropologies of nature, science and technology, religion, art, kinship, feminism, gender and sexuality, dance and film.

The roots of modern anthropology can be drawn back to ancient times, when some historians, geographers, and travellers became interested to know more about the origins, styles of life, behaviours, languages and cultures of human communities or societies living in other areas or regions of the world (Herodotus, 450-440BC; Zhang Qian, 138-125BC; GC Tacitus, AD89-93). Such kinds of investigation and study continued in medieval and Renaissance times (Xuan Zang, 640-660; Marco Polo, 1271-95; L de Varthema, 1502-10), and then further up to modern times (I Kant, 1798; KJV Rasmussen, 1921-24; J Malaurie, 1945-55).

Social and cultural anthropology started from the mid 19th century and continued by systematic researches until the present-day (LH Morgan, 1851-77; FU Boas, 1911-40; M Maus, 1924-34; ZN Hurston, 1926-37;

M Mead, 1928-70; RF Benedict, 1934-46; C Lévi-Strauss, 1958-78; CJ Geertz, 1960-83; ER Wolf, 1969-82; PE Farmer, 1999-2003).

As described in above chapter 3, the human lineage extended from hominids, in particular through the genus *Homo* and *Homo sapiens*. The process leading to modern humans and their evolution has been at least partially unveiled by scientific studies of physical anthropology, paleoanthropology, archaeology, ethology, neurology, linguistics, evolutionary psychology, embryology, genetics, and linguistics.

First members of *Homo* genus which moved from Africa to Eurasia were those belonging to *Homo erectus*, about 1.9 - 1.8 million years ago. They were followed by those belonging to a Homo species which also migrated out of Africa around 1.0 million years ago and descended as *Denisova hominid* found in southern Siberia, at Denisova Cave, and dated 41000 years ago, whose mitochondrial DNA (mtDNA) was distinct both from the Neanderthals and modern human (J Krause *et al.*, 2010). The migration from Africa to Eurasia continued by members of *Homo neanderthalensis* about 0.4 - 0.3 million years ago, which spread widely in the Near East, Europe and Siberia. At that time, 0.3 - 0.2 million years ago, members of *Homo rhodesiensis* were living in a large area consisting of South, East and North Africa.

Based on recently genetic discoveries, the common ancestor of living anatomically modern humans lived later than *Homo heidelbergensis* and the emergence of *Homo neanderthalensis*, in the East Africa as a "Y-chromosomal Adam", dated initially 338 thousand years ago (M Hammer, 2013) and finally around 200 thousand years ago.

The first members of *Homo sapiens* emerged in East Africa about 0.2 million years ago, being identified by their *Omo* remains discovered near the Omo River in south-western Ethiopia (REF Leakey, 1969), and later dated as 200 - 190 thousand-year old (J Fleagle *et al.*, 2008). They were followed by *Homo sapiens idaltu* (in Saho-Afar language *idaltu* 'elder, or first born'), living 160 - 154 thousand years ago, whose remains were discovered at Herto Bouri, near the Middle Awash site in Ethiopia's Afar Triangle (T White *et al.*, 2003). The earliest "X-mitochondrial Eve" was identified also in Africa and dated around 150 -120 thousand years, i.e. before the out of Africa migration.

Between 100 and 80 thousand years ago, the early members of Homo sapiens living in Africa differentiated into three main lines: 1st, bearing mitochondrial haplogroup LO (mtDNA) / A (Y-DNA), and colonizing Southern Africa as ancestors of the Khoisan people; 2nd, bearing haplogroup L1 (mtDNA) / B (Y-DNA), and settling Central and West Africa as ancestors of western pygmies; and 3rd, bearing haplogroups

L2, L3, and other mtDNA, and remaining in East Africa as ancestors of Niger, Congo, and Nilo-Saharan speaking peoples.

Modern humans' exodus from Africa took place on two different routes: one along the Nile Valley through the Isthmus of Suez to the Middle East, at least up to the present-day Israel, according to the 120 - 100 thousand years old human remains discovered at Jebel Qafzeh and Skhul sites; and another through the Bab-el-Mandeb Strait on the Red Sea (at that time with much lower level and narrower extension than today), to Arabian Peninsula, then to the Persian Gulf at least between 125 and 106 thousand years ago (A Lowler, 2011), and eventually into the Indian Subcontinent more than 75 thousand years ago. Meanwhile, coming from the Middle East, possibly through Central Asia, modern humans arrived in southern China, where their fossil remains were dated as early as 139 -111 thousands years ago (Wu Liu et al., 2010).

Furthermore, modern humans spread into South Asia 60 - 55 thousand years ago, and then colonized Indonesian islands between 50 and 40 thousand years ago, where they met previous inhabitants belonging to species *Homo floresiensis* identified by remains found in the island of Flores (P Brown et al., 2004), maybe interbreeding with them.

A branch of the modern humans from the Arabian Peninsula migrated through the East Mediterranean-Black Sea area into Europe between 45 and 43 thousand years ago, where coexisted and interbreed with previous Neanderthal inhabitants, giving rise to the *Cro-Magnon* race (L Lartet, 1868), as then genetically proved on the basis of 1 - 4% DNA inherited by the newcomers from the previous ones. This Neanderthal genetic contribution is a common characteristic for the most of non-African populations.

Modern humans' migration continued with colonization of Australia about 40 thousand years ago, East Asia 30 thousand years ago, Alaska and North America from 30 to 14 thousand years ago, and the Pacific islands of Polynesia between 1300BC and AD900.

Scientific studies of human capabilities and especially language were based on analyses of size, structure and function of brain, first showing that the cranial capacity varied from $1200...1800 \approx 1500$ cm^3 $\approx 1.50\cdot10^{-3}$ m^3 for *Homo neanderthalensis*, through $1400...1500 \approx 1450$ cm^3 $\approx 1.45\cdot10^{-3}$ m^3 for *Omo* and *Homo sapiens idaltu*, and $900...1800 \approx 1350$ cm^3 $\approx 1.35\cdot10^{-3}$ m^3 for *Cro-Magnon*, to $1000...1500 \approx 1250$ cm^3 $\approx 1.25\cdot10^{-3}$ m^3 for *present-day humans*. Many of the earlier anthropologists believed that the human level of evolution (including intelligence and communication) is related to the cranial capacity, but this opinion is not always confirmed and then the later anthropologists concluded that cranial capacity is not significant to establish the level of human evolution. This inconvenience can be overcome by re-conceptualizing the assessment of brain energy E [J = kg·m²·s⁻² = N·m = V·C = W·s] as a product of its extensive quantity or *capacity* q [m³] and its intensive quantity or *potential* p [J·m⁻³], where, as SI units, J = joule, kg = kilogram, m = metre, s = second, N = newton, V = volt, C = coulomb, and W = watt. The capacity q is simply the brain volume, while the potential p is complexly the brain efficiency of processing and storing information, speaking, imagining, intuiting, problem-solving, and so on. Using the abbreviations *Nt* for *Homo Neanderthalensis*, *Oi* for *Omo* and *Homo sapiens idaltu*, *Cm* for *Cro-Magnon*, *Ph* for *present-day humans*, their brain energy is expressed as:

$$E_{Nt} = p_{Nt} \cdot q_{Nt}, \; E_{Oi} = p_{Oi} \cdot q_{Oi}, \; E_{Cm} = p_{Cm} \cdot q_{Cm}, \text{ and } E_{Ph} = p_{Ph} \cdot q_{Ph}.$$

In the order imposed by evolution, the increase of brain energy can be displayed by the inequalities

$$E_{Nt} < E_{Oi} < E_{Cm} < E_{Ph}, \text{ or } p_{Nt} \cdot q_{Nt} < p_{Oi} \cdot q_{Oi} < p_{Cm} \cdot q_{Cm} < p_{Ph} \cdot q_{Ph},$$

whence

$$p_{Nt} < p_{Oi} \cdot (q_{Oi}/q_{Nt}) < p_{Cm} \cdot (q_{Cm}/q_{Nt}) < p_{Ph} \cdot (q_{Ph}/q_{Nt}),$$

and then

$$p_{Oi} > p_{Nt}/(q_{Oi}/q_{Nt}); \; p_{Cm} > p_{Nt}/(q_{Cm}/q_{Nt}); \; p_{Ph} > p_{Nt}/(q_{Ph}/q_{Nt}).$$

Based on these relationships, the evolution of brain energy and potential can be assessed knowing that:

(i) the average brain capacities $q_{Nt} \approx 1.50 \cdot 10^{-3} m^3$, $q_{Oi} \approx 1.45 \cdot 10^{-3} m^3$, $q_{Cm} \approx 1.35 \cdot 10^{-3} m^3$, $q_{Ph} \approx 1.25 \cdot 10^{-3} m^3$ (see above);

(ii) the Neanderthal's average brain energy $E_{Nt} \approx 4.748 \cdot 10^{10} J$ and potential (density of energy) $p_{Nt} \approx 3.165 \cdot 10^{13} J/m^3$ (see table in subchapter 3.3);

(iii) the present-day human's average brain power of around 26 watts (i.e. 26% of energy used by the entire body of an adult) and lifespan of about 70 years $\approx 2.209 \cdot 10^9$ seconds (according to recently published data), which result in an average energy $E_{Ph} \approx$ (26 watts)·(2.209·10⁹ seconds) $\approx 5.743 \cdot 10^{10} J$, and therefore an average potential $p_{Ph} = E_{Ph}/q_{Ph} \approx (5.743 \cdot 10^{10} J)/(1.25 \cdot 10^{-3} m^3) \approx 4.594 \cdot 10^{13} J/m^3$.

Furthermore, the chosen specimens' development of brain capacity, potential and energy can be displayed such as:

Specimen	Capacity q [$10^{-3}m^3$]	Ratio q/q_{Nt} [1]	Potential [$10^{13}J/m^3$]		Energy $E = p \cdot q$ [$10^{10}J$]
			referential $p_{Nt}/(q/q_{Nt})$	estimated $p > p_{Nt}/(q/q_{Nt})$	
Ph	1.25	0.833	3.800	4.594	5.743
Cm	1.35	0.900	3.517	4.095	5.528
Oi	1.45	0.967	3.273	3.625	5.256
Nt	1.50	1.000	3.165	3.165	4.748

Such a display of brain energy confirms the natural order of evolution: (*Homo neanderthalensis*) → (*Omo* and *Homo sapiens idaltu*) → (*Cro-Magnon*) → (*modern human*).

Human communication requires a sender, a message, and a recipient, and can extend from short to long times and distances, being completed once the receiver understands the sender's message. This communication with others involves three main steps: *thought* (information existing in sender's mind as a concept, idea, feeling, etc.); *encoding* (message sent to a receiver in words or other symbols), and *decoding* (receiver translating the words or symbols into an understandable concept or information). Verbal and non-verbal communication can be of various forms, such as body language, eye contact, sign language, haptic (by touch), mono- or polychronic, or media content (pictures, graphics, sound, and writing). In the case of disabled people, the communication includes the display of text, large print, tactile, Braille, accessible multimedia, written and plain language, human-reader, augmentative and alternative modes, means and formats (accessible information and communication technology). An effective communication implies *feedback* as a critical component.

Verbal communication, such as human spoken and pictorial languages, can be described as a *system of symbols* (sometimes known as *lexemes*) and the *grammar* (rules) by which the symbols are manipulated.

Oral communication, while primarily referring to spoken verbal communication, can also employ visual aids and non-verbal elements to support the conveyance of meaning. Oral communication include: speeches, presentations, discussions, and aspects of interpersonal communication.

Along with the communication through languages, the mass communication reached a large audience through media technologies including print, broadcast and internet media, covering the entire world. The development and perspective of worldwide communication (HM McLuhan, 1951-89) became the subject of the *communication theory* that is a field of information and mathematics that studies the *technical process of information* (CE Shannon, 2011) and the process of *human communication* (R Crag, 1999; S Littlejohn and K Foss, 2008).

Differences between human and animal communication mainly consist of:

(i) *Duality of structure of human language* - each human language has a fixed number of sound units, called *phonemes*, and these are combined to make *morphemes*; thus getting two levels of patterning, which is not prevailed in animal communication;

(ii) Humans use their linguistic resources to produce <u>new expressions and new sentences</u>, arranging and rearranging phonemes, morphemes, words and phrases to create new modes of expression, as so called *open-endedness of language*; while animal communication is a closed system that cannot produce new vocal signals to transmit novel events or experience;

(iii) Humans can talk about <u>real or imaginary situations, places and objects</u> far removed from their present surroundings and time, while animal communication is merely a response to stimuli from the immediate environment such as the presence of food or danger;

(iv) Human language is <u>*interchangeable*</u>, both men and women using the *same language interchangeably*; but the animal communication is often performed only by one sex of the sect, such as the bee dancing that is performed only by worker bees;

(v) Human language is <u>*culturally transmitted*</u>, and thus humans brought up in different culture acquire different language, but also, under the influence of other culture, can *learn another language*; while animals lack in this capacity, their communication ability being transmitted biologically and so restricted to its own inherited possibilities;

(vi) Language is a human <u>*symbolic system*</u>, being not only utterable but also displayable by *writing with certain symbols*, such as *alphabets*, so that ideas and events can be preserved and made known for future generations; by difference, animal communication fails to disclose in such a manner;

(vii) Human vocal cords can produce a <u>*large variety of sounds*</u>, each language making its own *selection in use of sounds*; while animals, including birds, have an entirely different biological structure for the formation of their sounds.

The ability to refer to things or states of being that are not in the immediate realm of the speaker is called <u>*high-level reference*</u>. There were identified six main aspects of the high-level reference systems: *theory of mind*; capacity to acquire *non-linguistic conceptual representations*, such as the object/kind distinction; *referential vocal signals*; *imitation* as a rational, intentional system; voluntary *control over signal production* as evidence of intentional communication; and *number representation* (T Matsuzawa, 1985; M Hauser *et al.*, 2002).

Language (from Latin *lingua* 'tongue, speech, language') is the human capacity for acquiring and using complex systems of communications, being usually defined as: human speech; variety of speech or body of words and idioms; mode of expression; diction; manner of expressing thought or feeling; artificial system of signs and symbols with rules for forming intelligible communications; as well as national branch of one of religious and military orders. In the human brain, language is derived and controlled from specialized centres of the cerebral cortex, in the left hemisphere for 97% of right-handed people and 19% of left-handed people, and in both left and right hemispheres for 68% of them. These centres are situated in five language areas of the brain, namely *Angular Gyrus, Supra-marginal Gyrus, Paul Broca's area, Carl Wernicke's area,* and *Primary Auditory Cortex,* which play their roles in processing and understanding language: the left hemisphere processing the linguistic meaning of prosody (rhythm, stress, and intonation of connected speech), while the right hemisphere processes the emotions conveyed by prosody.

The emergence of language is closely tied to the origin of modern human behaviour. There are numerous hypotheses about how, why, when, and where language might first have emerged. These hypotheses can be grouped either into the *'continuity theories'* or *'discontinuity theories'*. The first ones are based on the idea that language in its complexity could not simply appear from nothing in its final form, but evolved from earlier pre-linguistic systems of communication among humans' ancestors (JB Monboddo, 1773-92; I Ulbæk, 1998). The second ones are based on the opposite idea that language is a unique trait lacking among non-humans, and should suddenly appear during the human evolution only, for example by a single change mutation in one individual and instantaneously installed as a component of mind-brain in a nearly perfect form (AN Chomsky, 2004).

Linguists reject the term of 'primitive language' by at least two reasons: *(i)* no group of human beings today, even those living in a stone-age culture, speak what could be conceived of as a primitive language; and *(ii)* no known language in all of history was in any sense primitive.

The scientific study of language in its widest sense, in every aspect and in all its varieties is named <u>linguistics</u> (L Bloomfield, 1914-33; LT Hjelmslev, 1939-43; HH Hock and B Joseph, 1996; M Alinei, 1996-2000; A Dalby, 1998; G Price, 2000). Among the subfields of linguistics can be mentioned:

1) Study of language structure, or *grammar*, focusing on the system of rules followed by its speakers, which includes the study of morphology (the formation and composition of words), syntax (the formation and composition of phrases and sentences from words), and phonology (sound systems);

2) Study of language, concerning how languages employ logic structures and real-world references to convey, process, assign meaning, and how they resolve ambiguity, which comprises the study of semantics (how meaning is inferred from words and concepts), and pragmatics (how meaning is inferred from context);

3) Study of the broader context in which language is influenced by social, cultural, historical and political factors, including evolutionary linguistics (related to the origins and growth of languages), historical linguistics (related to language change), sociolinguistics (referring to the relation between linguistic variation and social structures), psycholinguistics (exploring the representation and function of language processing in brain), language acquisition (related to the acquirement of language by children and adults), and discourse analysis (involving the structure of text or conversations).

Recently, the new enunciated *language organ* treats human language as the manifestation of mind faculty whose nature is determined by human biology, and whose functional properties can be explored just as physiology explores the functional properties of physical organs (S Anderson and D Lightfoot, 2002).

Language is the most precious treasure of every culture, preserving and transmitting traces of its historical background. For instance, the present-day English includes elements from pre-Celtic, Celtic, Latin, German, Scandinavian, and French, as the result of historical and cultural development. As media's influence is spreading, there is an increasing tendency towards a uniformity of our cultures, but meanwhile a threat for many languages with extinction. Like their speakers, many languages were emerging, differentiating, flourishing, and spreading, while other languages were diminishing, surviving, and even extinguishing. The attempt to create an international language, such as *Esperando* intended to promote world peace (LL Zamenhof, 1893-1905), failed maybe because its discordance with the natural course of language development.

Early human language development can by roughly reconstructed on the basis of children's pre-linguistic communication such as:

0 - 2 months - reflexive crying, vegetative sounds (coughs, sneezes), and sounds reflecting physical state;

2 - 5 months - cooing and laughter, early consonants, sounds from the back of throat, laughs and giggles;

4 - 6 months - vocal play, babbling gets, range and pitch play, and bilabial trills;

6 - 12 months - reduplicated babbling (e.g. 'mamma'), pitch control,

ability to sound out some consonants and vowels;

9 - 18 months - non-reduplicative babbling, varying of consonants and vowels.

From the birth to three years, the children's speech and language developmental milestones are considered the following:

Birth - 9 months - the pre-linguistic stage,

9 - 18 months - beginning of semantics and syntax,

18 - 24 months - development of semantics and syntax,

After 2 - 3 years - complex syntax and fulfilment of language.

Linguistics is based on a series of observations, studies, and analyses on the differentiation, changes, rules, and relationship of various languages, as well as the comparison to each another, decipherment of former or extinct languages, research for cerebral centres of language, and language programming (A Alexandrescu, 2001-04). It includes the following subdisciplines: *descriptive linguistics* (grammar of single language), *theoretical linguistics* (conceptualizing and defining the nature of language), *sociolinguistics* (manner in which languages are used for social purposes), *neurolinguistics* (manner in which language is processed in the human brain), *computational linguistics* (use of computers for modelling languages), and *historical linguistics* (tracing individual history of a language and reconstructing trees of language families by comparative methods).

4.2. Distinguishing particularities

A human is distinctive from all predecessors by four major particularities: anatomical modifications, genetic changes, cerebral structure and function, social competence and conditions.

Anatomical modifications
Although Neanderthals had a simple kind of speech, the human language emerged from very beginning of Homo sapiens, i.e. about 200000 years ago, later evolved by writings and scripts, literature and printing, and will have the same fate as their speakers.

Looking backwards, and comparing with non-human primates, the humans are characterized by significant adaptations, such as bipedalism, increased brain size, lengthened ontology (gestation and infancy) and decreased sexual dimorphism. One important morphological change included the evolution of a power and precision grip.

The speech organs evolved in the first instance not for speech, but for more basic bodily functions such as feeding and breathing. Our species' unprecedented use of the *tongue, lips,* and *vocal organs* as instrument of communication developed in association with *larynx* (voice box), *respiratory control*, and a special shaped lingual bone, named *hyoid bone* (Latin *os hyoideum*, from Greek χιοειδες 'upsilon shaped') that is situated in the anterior midline of neck between the chin and the thyroid cartilage, where it plays a crucial role in speaking by supporting the root of the tongue. Recent discoveries and analyses indicate that the hyoid bone existed even in *Homo heidelbergensis* from over 500000 years ago, such a bone being found in Spain. A 60000-year-old Neanderthal's fossilized hyoid bone discovered in the Kebara Cave in Israel has a horseshoe-shaped form that looks and worked in a very similar manner to a modern human's one, but not placed in the right position. Furthermore, psycholinguistic researches suggest that Neanderthals shared with modern humans a similar capability of speech and language (J Steele *et al.*, 2013; D Deliu and S Lavinson, 2013).

The *phonatory process*, or voicing, takes place when air is expelled from the lungs through *glottis*, creating a pressure drop across the larynx. When this drop becomes sufficiently large, the vocal folds start to oscillate. The minimum pressure drop required to achieve phonation is called *phonation threshold pressure*, which for humans with normal vocal folds is approximately 2-3 centimetres column of water. The sound produced by larynx is a harmonic series, i.e. consists of a fundamental tone (called the fundamental frequency, as the main acoustic cue for the percept pitch) accompanied by harmonic overtones, which are multiples of the fundamental frequency. According to the source-filter theory, the resulting sound excites the resonance chamber that is the *vocal tract* to produce the individual speech sounds. The *voice* consists of sound made by a human using the *vocal folds* for talking, singing, laughing, crying, screaming, etc. Habitual speech fundamental frequency ranges between 75 and 150 hertz for man and between 150 and 300 hertz for woman. The mechanism for generating the human voice can be subdivided into three parts: the lungs, the vocal folds within larynx, and articulators.

Genetic changes

Human evolutionary genetics studies how one human genome differs from the other, the evolutionary past giving rise to it, and its current effects. Differences between genomes have anthropological, medical and forensic implications and applications. Moreover, genetic data can provide important insight into human evolution.

Researches using the molecular clock indicate that gorillas split up 8.4 - 6.2 million years ago, then chimpanzees and humans split up 6.2 - 4.6 million years ago. The loss of the sarcometric myosin gene MYH16 in the human lineage led to smaller masticatory muscles, and the mutation that led to the inactivation (a two base pair deletion) occurred just before 2.4 million years ago, predating the appearance of series *Homo habilis-erectus-ergaster* in Africa. The following period was marked by a strong increase in cranial capacity (H Stedman *et al.*, 2004).

The human gametes comprise 23 pairs of chromosomes, the number of possible combinations of chromosomes being 2^{23} = 8,388,608. Human gametes normally end up with 23 chromosomes, but the origin of any particular one is randomly selected from paternal or maternal chromosomes, so contributing to the genetic variability of progeny.

The human genome assembles all DNA that a person possesses, being made up of our chromosomes and also present in our mitochondria. From the human genome, stored on 23 chromosome-pairs and in the small mitochondrial DNA, 22 belong to autosomal chromosome-pairs, while the remaining pair is sex determinative. The haploid (with reduced number of chromosomes) human genome consists of approximately 3 billion DNA base pairs representing about 30 thousand genes, and contains around 23 thousand protein-coding genes, but actually only about 1.5% of the genome codes for proteins, the rest consisting of non-coding RNA genes, regulatory sequences, introns, and non-coding DNA.

The X chromosomes of humans and chimpanzees appear to have diverged about 1.2 million years later than the other chromosomes. Humans have 23 pairs of chromosomes, while chimpanzees, gorillas, and orang-utans have 24. This difference is because the human chromosome 2 is a fusion of two chromosomes 2a and 2b that remained separate in the other primates.

Our closest non-human relatives, chimpanzees, are unable to speak even after years of intensive training. This is because they have not the two functioning copies of the gene called FOXP2 needed for speech and language (C Lai *et al.*, 2001). This gene is encoded for production of the forkhead box protein P2 that is involved in human speech and language (also in bird song) and played an important role in human evolution.

Cerebral structure and function

Neurons are the basic information processing structures in the central nervous system. The function of a neuron is to receive input 'information' from other neurons. The neurons are connected by synapses through which 'information' flows from one neuron to another. Hence, they process all of the 'information' that flows within, to, or out of the central nervous system, as *motor information* through which animals are able to move; *sensory information* through which animals are able to see, hear, smell, taste, and touch; and *cognitive information* through which we are able to reason, think, dream, plan, remember, and to do everything else that we do with our mind.

The support to the neurons is provided by glia (or glial cells), which also attend to the brain's various house-keeping functions, such as removing debris after neuronal death, allowing networks of neurons to remain connected.

Usually, a neuron has four distinct parts: *cell body* (soma), *dendrites* which receive incoming signals from other neurons; *axon* which allows the flow of outgoing signal to other neurons; and *axon terminals* which contain neurotransmitters (a chemical medium through which signals flow from one neuron to the next at chemical synapses).

Neurons have capabilities for *intracellular signalling* (communication within the cell), and *intercellular signalling* (communication between cells), showing that they have evolved special functions for sending *electrical signals* (action potentials) along axons. This mechanism, called *conduction*, indicates how the cell body of a neuron communicates with its own terminals via the axon. The communication between neurons is achieved at synapses by the process of *neurotransmission.*

Electrical properties of neurons are controlled by a wide variety of biochemical and metabolic processes, most notably the interactions between neurotransmitters and receptors that take place at *synapses* (US von Euler, 1940-50; AL Hodgkin and AF Huxley, 1948-50; B Katz, 1945-70; JC Eccles, 1951-63).

The cerebral hemispheres form the largest part of the human brain and are situated above most other brain structures. They are covered with the cortical layer of *neocortex*, also called the *neopallium*, or *isocortex*, with six component layers, and deep grooves called *sulci*, and wrinkles called *gyri*, which consist of grey matter, or neuronal cell bodies and unmyelinated fibres, surrounding the deeper white matter (myelinated axons). The neocortex is involved in higher functions, such as sensory perception, generation of motor commands, spatial reasoning, conscious thought, working memory, speech, and language, as well as social and emotional processing (I Kant, 1785; WR Hess and ACE Moniz, 1945-49; RW Sperry, 1955-65; AG Gilman and M Rodbell, 1981-92). Underneath the cerebrum lies the *brainstem*, resembling a stalk to which the cerebrum is attached, and controlling the breath, heart rate, and other automatic processes. At the rear of the brain, beneath the cerebrum and behind the brainstem, is the *cerebellum*, a structure with a horizontally furrowed surface, which is responsible for the body's balance, posture, and coordination of movement.

The two hemispheres consist of four lobes, namely *frontal* (involved in reward, attention, short-term memory tasks, planning, and motivation), *parietal* (containing areas involved in somatosensation, hearing, language, attention, and spatial cognition), *occipital* (including a small area dedicated to vision), and *temporal* (involved in auditory perception). Each hemisphere of the brain interacts primarily with one half of the body, but the connections are crossed, i.e. the left hemisphere interacts with the right side of the body, and the right hemisphere interacts with the left side of the body.

Within the adult human brain there are $(8.6...12.0) \cdot 10^{10}$ neuronal cells (neurons) of which $(1.63...2.27) \cdot 10^{10}$ are in the *cerebral cortex*, and $(6.9...9.63) \cdot 10^{10}$ in the *cerebellum*, as well as an approximately equal number of non-neuronal cells called *glial cells*. Each neuron is connected to other neurons by $(1...10) \cdot 10^3$ *synapses*, so that the total number of synapses is $8.6 \cdot 10^{13}...1.2 \cdot 10^{15}$. There are about 10,000 specific types of neurons, which can be grouped into three kinds: *motor neurons*, *sensory neurons*, and *interneurons*, conveying motor information, sensory information, and information between different types of neurons respectively (CS Sherrington and ED Adrian, 1919-

30; J Erlanger and HS Gasser, 1921-30; HH Dale and O Loewi, 1928-35; B Katz, 1945-70; JC Eccles, 1951-63).

The neurons in the cerebral cortex are located in *neural membranes* that have an electric potential of minus $(5...7)\cdot10^{-2}$ volts, capacitance of $1\cdot10^{-10}$ farads per square metre, and charge across the membrane of about 1 coulomb per volt per cubic metre. The signals from neurons propagate along *axons* with speeds up to 150 metres per second, ending at the axon terminals where neurotransmission takes place by *electrical synapses* (dendrite-dendrite), the output of which consists of electrical signals themselves, or *chemical synapses* (axon-dendrite), the output of which consists of neurotransmitters. Electrical neurotransmission is communication between two neurons at the electrical synapses. Chemical neurotransmission occurs at chemical synapses to overcome the 'short' in an electrical circuit, thus acting like a chemical messenger, linking the action potential of one neuron with a synaptic potential in another.

In the adult human brain, the myelinated axons extend up to about 1.5 kilometres for a woman and 1.8 kilometres for a man, the neuronal potential can be up to $1\cdot10^{-9}$ amperes, and the postsynaptic potential can range from $4\cdot10^{-4}$ to $2\cdot10^{-2}$ volts.

80-85% of the brain's neurons are housed by the *cerebral cortex*, being arranged in six layers with a total thickness of $(1.5...4.5)\cdot10^{-3}$ metres on a surface area increasing from 0.06 m^2 at birth to 0.25 m^2 at adulthood, in a half million neocortical columns of about 0.5 millimetres diameter and 2 millimetres depth, each containing around 60,000 neurons. Nevertheless, from the total large number of neurons, only between 0.01% and 0.10% is properly used, the difference remaining available for further activation. Therefore, from the total mental area ranging from $6\cdot10^{-2}$ to $2.5\cdot10^{-1}$ square metres only between $6\cdot10^{-6}$ and $2.5\cdot10^{-4}$ square metres is active for cerebral functions, including language.

By function, the left cerebral hemisphere provides logic, while the right cerebral hemisphere provides creativity and originality. These two cerebral hemispheres represent circa 70% of the brain, and their surface area is increased 30 times by the *sulci* (fissures between convolutions of brain) and *gyri* (convolutions of the brain). Nerve fibres cross over in the brain stem so that the right hemisphere is linked with the left side of the body and vice versa. The left hemisphere controls speech and logic, but relies on the right side to make up three-dimensional and artistic representations. Other parts of the forebrain are the frontal lobe controlling movement; the prefrontal area concerned with intelligence and personality; the parietal lobe for sensation and body position; the smaller occipital lobe for vision; and the temporal lobe for hearing.

Visual perception has some hierarchical character in the recognition of visual objects, which takes place in the temporal lobe. The determination of location in space takes place in the parietal lobe of the cortex. About one-quarter of the cerebral cortex in humans is dedicated to processing visual information.

Memory is distributed, not localized, for example: recent memory is processed by the hippocampus; long-term memory is processed in the parietal lobe and in the temporal lobe; and recalling memory activates the frontal lobe.

The global control of *attention* apparently comes from the prefrontal cortex. In attention, a crucial role is played by the pulvinar (the largest nucleus in the thalamus), occupying two-fifth of the thalamic volume.

The centre of the *fight/flight response* is in the periqueductal gray, also called central grey, which surrounds the cerebral aqueduct of the midbrain (just below the tectum).

In human brains, for right-handed people, the left hemisphere seems to play a greater role in controlling verbal and some mathematical skills, whereas the right hemisphere is more involved in spatial perception, and in reverse for left-handed people. Each personality is based on characteristics genetically inherited and psychically acquired as a combined result of environmental, emotional, educational, and social conditions. The human mind is creative mainly by *imagination*, the capacity to form images by taking things, facts, and events together; and *intuition*, which is the aptitude to perceive links between the things, facts, and events by correlating some of them. Furthermore, the formed images and perceived links result in structures and chains respectively, forming a basis for technology, science, religion, civilization, justice, language, art, philosophy, etc. in particular, and for knowledge in general (JA Comenius, 1630-31; ÉB de Condillac, 1746-54; A Schopenhauer, 1819-41). A standard of mental integrity involves honesty, truthfulness, kindness, love and understanding, patience, tolerance, perseverance, logic, and sincerity.

During the development of the human neural system, there are six main stages of improvement in function: *(i)* direct reaction to a sensor signal; *(ii)* instinctive response to an inherited pattern which is associated with danger or food; *(iii)* development of sensory memory and comparison; *(iv)* ability to imagine, to mentally construct sensor patterns, to remember them, and then to use them as if they were real in the value summation neural circuits, and it provides a creativity element in the instinctive value summation process; *(v)* conscious thought, an awareness of identity, a feeling of personal management; and *(vi)* intelligent thought, a rigid methodology, and a mostly painful process.

Therefore, the mental energy, or psychic activity (GT Fechner, 1836-76; WB Carpenter, 1846-88; G Boole, 1847-54; WM Wundt, 1863-1920; GF Stout, 1891-1931; S Freud, 1900-23; HJ Eysenck, 1953-90), derives from the *cerebral cortex* (neocortex) that, during the human lifetime, has an active volume varying between $9.0 \cdot 10^{-9}$ and $1.12 \cdot 10^{-6}$ cubic metres, and an average power requirement of 26 watts.

Neurotransmission, or synaptic transmission, is the process by which neurotransmitters (signalling molecules) are released by a presynaptic neuron, binding to and activating the receptors of a postsynaptic neuron. This process usually takes place at a synapse, and occurs when an action potential is initiated in a presynaptic neuron. The binding of neurotransmitters to receptors in the postsynaptic neuron can trigger either short term changes, such as changes in the membrane potential called postsynaptic potentials, or longer term changes, by the activation of signalling cascades. Nerve impulses play an essential role in the propagation of signals, which are sent to and from the central nervous system via efferent and afferent neurons in order to coordinate skeletal and cardiac muscles, bodily secretions and organ functions critical for the long-term survival of vertebrate organisms such as mammals.

Neurons form networks through which nerve impulses travel, each neuron receiving as many as 15,000 connections from other neurons. Except an electrical synapse through a gap junction, neurons do not touch each other, having synapses as contact points. A neuron transports its information by a nerve impulse, and when a nerve impulse arrives at the synapse, it releases neurotransmitters, which influence another cell, either in an inhibitory or an excitatory way. The next neuron may be connected to many other neurons, and if the total of excitatory influences is more than the inhibitory influences, it will also 'fire', i.e. it creates a new action potential at its axon hillock, thus passing on the information to yet another neuron, or resulting in an experience or an action.

At synapses, the following *stages in neurotransmission* occur: 1st, synthesis of the neurotransmitter, which can take place in the cell body, in the axon, or in the axon terminal; 2nd, storage of the neurotransmitter in storage granules or vesicles of the axon terminal; 3rd, penetration of calcium within the axon terminal during an action potential, causing release of the neurotransmitter into the synaptic cleft; 4th, the transmitter binds to and activates a receptor in the postsynaptic membrane; and 5th, deactivation of the neurotransmitter, that is either enzymatically destroyed, or taken back into the terminal from which it came, where it can be reused, or degraded and removed.

Each neuron connects with many other neurons, receiving numerous impulses from them, these impulses being added together at the axon hillock in a process called *summation*. If the neuron only gets excitatory impulses, then it also generates an action potential; but if, instead, the neuron gets as many inhibitory as excitatory impulses, then the inhibition cancels out the excitation and the nerve impulse stops there.

There are two kinds of summation: *spatial summation*, meaning that the effects of impulses received at different places on the neuron add up, and that the neuron may fire when such impulses are received simultaneously, even if each impulse on its own would not be sufficient to cause firing; and *temporal summation*, meaning that the effects of impulses received at the same place can add up if the impulses are received in close temporal succession, thus the neuron may fire when multiple impulses are received, even if each impulse on its own would not be sufficient to cause firing.

Neurotransmission implies both *convergence* and *divergence* of information. First, one neuron is influenced by many others, resulting in a convergence of input, and when the neuron fires, the signal is sent to many other neurons, resulting in a divergence of output; many other neurons being influenced by this neuron.

Cotransmission is the release of several types of neurotransmitters from a single nerve terminal. At the nerve terminal, neurotransmitters are present within 35-50 nanometre-wide membrane-encased vesicles called synaptic vesicles. Release of neurotransmitters implies that synaptic vesicles transiently dock and fuse at the base of specialised 10-15 nanometre-diameter cup-shaped lipoprotein structures at the presynaptic membrane called porosomes. The structure of the neuronal porosome (proteome) has been solved, providing the molecular architecture and the complete composition of the machinery.

Storage is one of three core processes of memory, along with recall and encoding. Storage in neural networks is described as follows: *netlet* - a small, functionally distinct collection of cells within a layer network, corresponding to single memory location in a neural storage system; *input pathway* - a pathway originating at the sensory microelectrode implanted in each axon; *initiate storage* - effectual input that causes the information to be stored; *initiate recall* - command input that causes stored information to be played in output; *storage* - as storage that occurs in real time, and as network that stores the spatial, time varying firing pattern arriving along the input neurons; and *consolidation* - the internal process of modifying the coupling parameters that effect storage.

The brain is the primary control centre of the body, a highly sophisticated complex organ containing billions of nerves that can simultaneously process information from bodies, operate internal organs, control movement, generate emotions and thoughts, store and recall memories. Observations and studies on brain functioning started from ancient times (C Galenus, AD170-190), continued during the Renaissance (J Peyligk, 1499; A Vesalius, 1543), in the 19th century (JE Purkinje, 1837; JM Charcot, 1870-90; W Macewen, 1879) and the 20th century (HW Cushing, 1909; ACE Moniz, 1920-40), until the present time (RH Hamilton, 2011).

In psychology, research related to neural information processing included *experiments in perception, problem-solving in animals*, and *Gestalt psychology* (W Köhler, 1913-20). The word *Gestalt*, meaning in German 'essence or shape of an entity's complete form', was used for this kind of psychology to provide the foundation for the modern study of perception, emphasizing that the whole of anything is greater than its parts, i.e. the attributes of the whole are not deducible from analysis of the parts in isolation.

Recent studies in neurology focused on *neural networks* (C Williams *et al.*, 1995), and *models of neural systems* (P Dayan and L Abbott, 2001) involving *information processing*. Subsequently, there were created *neural networks*, as simplified models of the brain composed of large numbers of units (analogous to neurons) together with weights that measure the strength of connections between units. Experiments on models of this kind have demonstrated an ability to learn such skills as face recognition, reading, and detection of simple grammatical structure.

The transmission of information is accomplished in two ways, electrically and chemically, common neurotransmitters including: *acetylcholine, dopamine, norepinephrine* (noradrenaline), and *serotonin*. In order to understand the brain functioning, the first step is to study the mechanisms of neurotransmission and storage of information within the brain.

In the human brain, there are specific centres and structures which are differentiated for controlling forms of expression such as emotion and language.

Emotion was initially suggested to be related to a group of structures in the centre of brain, called the *limbic system*, which mainly includes the *hypothalamus, cingulate cortex*, and *hippocampi* (P Broca, 1861-78);

but currently is thought to be most associated with the *amygdala* (anterior to the hippocampi near temporal poles, engaging the neural machinery that results in negative emotions, especially fear, anxiety and surprise); *prefrontal cortex* (the very front of the brain, behind the forehead and above the eyes, being competent in foresight, planning and self-control); *anterior cingulate* (in the middle of the brain, just behind the prefrontal cortex, playing a role in attention, as well as conscious, subjective emotional awareness and initiation of motivated behaviour); *ventral striatum* (with a role in emotion and behaviour, a part of it called 'nucleus accumbens' being involved in the experience of goal-directed positive emotion, namely pleasure and even addiction); and *insula* (playing a role in the bodily experience of emotion, and in experiencing the emotion of disgust). The hypothalamus is the source of many of the most elementary emotions, such as hunger, thirst, chills, and ultimately pleasure and pain.

The *use of language* is controlled from specialized centres of the cerebral cortex, which are situated in the five areas mentioned above, namely *Angular Gyrus*, *Supramarginal Gyrus*, *Broca's area*, *Wernicke's area*, and *Primary Auditory Cortex* (P Broca, 1861-78; C Wernicke, 1874-85; KL Pike, 1943-60). The centres within these areas for language are interconnected within a *total area* varying during the life time between 0.01 and 0.04 square metres, where the neurons are properly active between 0.01% and 0.10%, and therefore the *active area* for language ranges between $1 \cdot 10^{-6}$ and $4 \cdot 10^{-5}$ square metres. This active area multiplied by the *thickness* of cerebral cortex, that varies between $1.5 \cdot 10^{-3}$ and $4.5 \cdot 10^{-3}$ metres, results in an linguistic *active volume* ranging from $(1 \cdot 10^{-6}) \cdot (1.5 \cdot 10^{-3}) = 1.5 \cdot 10^{-9}$ to $(4 \cdot 10^{-5}) \cdot (4.5 \cdot 10^{-3}) = 1.8 \cdot 10^{-7}$ cubic metres.

The linguistic signals are propagating in neocortex with a *speed* decreasing during the life time approximately from 150 to 23 metres per second. As the mental active volume, ranging from $9.0 \cdot 10^{-9}$ to $1.12 \cdot 10^{-6}$ cubic metres requires a power of around 26 watts; it follows that the linguistic active volume, ranging from $1.5 \cdot 10^{-9}$ to $1.8 \cdot 10^{-7}$ cubic metres, requires a *linguistic power* varying from 4.17 to 5.63 watts, i.e. between 16 and 22% from the total brain power (M Albu, 2013).

Social competence and conditions

Social competence is a complex concept consisting of social, emotional, cognitive and behavioural skills, as well as motivation and expectancy sets, needed for successful social adaptation, including social and interpersonal communication. Among the factors contributing to social competence can be mentioned: *temperament* (M

Semrud-Clikeman, 2007), *attachment* (M Ainsworth, 1979), and *parenting style* (D Baumrind, 1991).

Social structures may play certain roles in *language emergence and evolution* (Tao Gong and W Wang, 2005), such as language emergence under initialized, unchanged social structure (as a structure with popular agent, popularity rate, acceleration/deceleration effect, with tendencies towards language divergence or convergence, language change, and other socio-economic factors), and co-evolution of language and social structure based on the natural understanding when using an evolving language implying successful communication, linguistic friendship, linguistic popularity, and local-view assumptions). All human societies have in common underlying features related to a certain level of cooperation and trust, below which even simple forms of language could not emerge. Differently from apes, humans have a higher level of cooperation consisting of mutual interpenetration of minds, and resulting from their *social conditions* in connection with psychological evolvement for adaptation to a particular style of life, based on hunting and gathering. Those evolving humans compensated their vulnerability to dangerous predators by an unprecedented development of social cooperation, material culture and strategies for remembering, transmitting and exchanging accumulated knowledge (C Knight and C Power, 2007). The groups of hunter-gatherers established a sort of egalitarianism, for example only approving a form of violence ritually organized by their community that in fact enforced an egalitarian law (C Boehm, 2001). When the human groups increased in size, the communities developed a social intelligence, as the ability to negotiate alliances, in turn driving selection pressures for neocortical expansion (RIM Dunbar, 1996).

The study of social organization in non-state societies is now called *social anthropology*.

Linguistic anthropology is the interdisciplinary study of how language influences social life, representing a branch of anthropology, originating from the endeavour to document endangered languages, and encompassing almost any aspect of language structure and use (A Duranti, 2004). Linguistic anthropology explores how language shapes communication, forms social identity and group membership, organizes large-scale cultural beliefs and ideologies, and develops a common cultural representation of natural and social worlds.

A distinct field of linguistics is *ethnolinguistics* which studies the relationship between language and culture, as well as the way different ethnic groups perceive the world.

Sociolinguistics studies the consequences of social aspects, including cultural norms, expectations, and context, on the way language is used, and the effects of language use on society.

As a relatively new branch, *computational linguistics* is an interdisciplinary field concerned with the statistical or rule-based modelling of natural language from a computational perspective; in which interdisciplinary teams include linguists, language experts, computer scientists, experts in artificial intelligence, mathematicians, logicians, philosophers, as well as cognitive scientists, psychologists, psycholinguists, anthropologists and neuroscientists. Computational linguistics is related to the new field of study for developing algorithms and software to intelligently process the language data; and comprises several kinds of approach, such as: *(i) developmental approach* for predicting future linguistic evolution, and giving insight into the evolutionary history of modern languages; *(ii) structural approach* for better understanding how the language works on a structural level, and for further probing the structure of human discourse; *(iii) production approach* for allowing the responses to written statements and questions posed by a user, and for producing language in a more naturalistic manner; and *(iv) comprehension approach* for allowing natural language processing, and interpreting naturally written commands within a simple rule governed environment (D Jurafsky and J Martin, 2009; F Mairesse and M Walker, 2011).

4.3. Lineage timing

There are several *hypotheses for language origin*, such as: *(i)* 'mother tongue', according to the principle of 'kin selection'; *(ii)* 'obligatory reciprocal altruism', expressing the principle 'if you scratch my back, I'll scratch yours'; *(iii)* 'gossip and grooming"; *(iv)* 'ritual-speech co-evolution'; or *(v)* 'Tower of Babel', in connection with the biblical story. Humans communicate with each other by means of: *voice* - the sound made when air from lungs is pushed between vocal folds in larynx, causing them to vibrate; *speech* - talking as one way to express language, involving the precisely coordinated muscle actions of the tongue, lips, jaw, and vocal tract to produce recognizable sounds that make up language; and *language* - a set of shared rules that allow people to express their ideas in a meaningful way, not only verbally but also by writing, singing, or making other gestures, such as eye blinking or mouth movements. A language consists of four main components: *phonology* - structure and sequence of speech sounds; *semantics* - vocabulary and manner in which concepts are expressed through words; *grammar* - *syntax* as rules in which words are arranged into sentences, and *morphology* as use of grammatical markers such as tense, active or passive voice, affixes etc.; and *pragmatics* - rules for appropriate and effective communication, by use of language for greeting, demanding, etc., change of language for talking differently, and practices such as turn-talking, or staying on topic.

The *early humans* were using forms of *protolanguage* that lack: *(i)* a fully developed syntax; *(ii)* tense, aspect, auxiliary verbs, etc.; and *(iii)* a non-lexical vocabulary (D Bickerton, 2009). According to available data regarding the origin and evolution of humans (see subchapter 4.1) and recent studies on proto-language reconstruction (D Bickerton, 2002; M Arbib and D Bickerton, 2010), proto-language lineage can be marked by several stages succeeding, in thousand years ago, such as:

240 - 230: *Proto-human language*, the most recent common ancestor of all world's languages, presupposing *monogenesis* of all known languages (except pidgins, creoles, and sign languages), used by *Homo neanderthalensis* in Europe, the Middle East and Siberia; and *Homo rhodesiensis* in South, East and North Africa;

200 - 190: *Earlier African proto-language*, spoken by the *Homo sapiens*, called *Omo*, in Ethiopia;

160 - 150: *Later African proto-language*, spoken by *Homo sapiens* *idaltu* in Ethiopia's Afar Triangle;

120 - 115: *Early Asian proto-languages*, spoken by African descendants, in the Middle East and southern China;

80 - 75: *South and south-east Asian proto-languages*, in India and Indo-China; *Indonesian proto-languages*, in Indonesian islands; and *European proto-languages*, from South-east to West of Europe;

40 - 38: *Aboriginal proto-language*, in Australia; *East and North-east Asian proto-languages*, from China to Chukotka Peninsula; *North American proto-languages*, from Alaska to Rio Grande; *Mesoamerican and South American proto-languages*, from Rio Grande to Tierra del Fuego; and *Pacific proto-languages*, in islands of the Pacific Ocean.

Therefore, a preliminary timeline of proto-languages comprises the following sequences delimited by approximately transitional times, in million years ago

\ 0.235 \ Proto-human

\ 0.195 \ Earlier African

\ 0.155 \ Later African

\ 0.117 \ Early Asian

\ 0.078 \ South and South-east Asian, Indonesian, European

\ 0.039 \ Australian, North-east Asian, American, Pacific

Referring to the origin of life c.4200 million years ago, the transitional times, also in million years, become:
4200 - 0.235 = 4199.765 \ 4200 - 0.195 = 4199.805 \
4200 - 0.155 = 4199.883 \ 4200 - 0.117 = 4199.883 \
4200 - 0.078 = 4199.922 \ 4200 - 0.039 = 4199.961 \ the present time.

In the hominids' timeline (see table at the end of subchapter 3.3), the latest $N/10^3$ = γ sequence is delimited by values of argument \1.180\...\1.881\ corresponding to *Homo sapiens* with his languages. This latest sequence can be detailed by $N/10^4 = \delta$ sequences which are suitable for displaying the succession of proto-languages.

Remembering that the final time is $t_\bullet$ = 4400 million years from the origin of life, the concluding transitional times t_δ = $t_\bullet \cdot tanh(\delta)$ are calculated, in million years from the same origin, as

$t_{1.8800} = t_\bullet \cdot tanh(1.8800) \approx (4400) \cdot (0.9544921) \approx 4199.7653;$

$t_{1.8801} = t_\bullet \cdot tanh(1.8801) \approx (4400) \cdot (0.9545010) \approx 4199.8044;$

$t_{1.8802} = t_\bullet \cdot tanh(1.8802) \approx (4400) \cdot (0.9545099) \approx 4199.8436;$

$t_{1.8803} = t_\bullet \cdot tanh(1.8803) \approx (4400) \cdot (0.9545188) \approx 4199.8827;$

$t_{1.8804} = t_\bullet \cdot tanh(1.8804) \approx (4400) \cdot (0.9545277) \approx 4199.9218;$

$$t_{1.8805} = t_\bullet \cdot tanh(1.8805) \approx (4400) \cdot (0.9545366) \approx 4199.9609;$$
$$t_{1.8806} = t_\bullet \cdot tanh(1.8806) \approx (4400) \cdot (0.9545454) \approx 4200.0000.$$

By these transitional times, the timeline of proto-languages is integrated in the life timeline, as shown in the table below.

$N/10^4$ = δ	Time (million years)		Sequences of proto-languages
	from origin t_δ = $t_\bullet \cdot tanh(\delta)$	from present t_δ - 4200	
	4400	+200	
...	...	...	
1.8806	4200.0000	±0	_
1.8805	4199.9609	-0.0391	_Australian, NE Asian, American, Pacific
1.8804	4199.9218	-0.0782	_S and SE Asian, Indonesian, European
1.8803	4199.8827	-0.1173	_Early Asian
1.8802	4199.8436	-0.1564	_Later African
1.8801	4199.8044	-0.1956	_Earlier African
1.8800	4199.7653	-0.2347	_Proto-human
...	...	...	
0	0	-4200	

The major proto-languages are classified such as:

Proto-human language and *Proto-Dené-Caucasian language* are reconstructed as macrofamilies of proto-languages;

Proto-Afroasiatic languages consist of *Proto-Semitic*; and *Proto-Berber*;

Near East proto-languages include *Proto-Anatolian*; *Proto-Northwest Caucasian* (Proto-Abazgi, Proto-Circassian); and *Proto-Kartvelian*;

Sino-Tibetan proto-languages consist of *Proto-Chinese* (Proto-Mandarin, Proto-Wu); and *Proto-Tibeto-Burman*;

Eurasian proto-languages comprise *Proto-Basque*; *Proto-Dravidian*; and *Proto-Indo-European* (Proto-Greek, Proto-Armenian, Proto-Indo-Iranian, Proto-Balto-Slavic, Proto-Celtic, Proto-Germanic, and Proto-Romance);

North Asian proto-languages include *Proto-Turkic*; *Proto-Mongolic*; *Proto-Uralic*; and *Proto-Chukotko-Kamchatkan*;

American proto-languages consist of *Proto-Eskimo-Aleut*; *Proto-Algonquian*; *Proto-Iroquoian*; *Proto-Uto-Aztecan*; *Proto-Mayan*; and *Proto-Oto-Manguean*;

Pacific Rim proto-languages comprise *Proto-Pama-Nyungan*; *Proto-Austronesian* (Proto-Malyo-Polynesian, Proto-Oceanic); *Proto-Tai-Kadai* (Proto-Northern Tai-Kadai, Proto-Southern Tai-Kadai); *Proto-Hmong-Mien*; and *Proto-Mon-Khmer*.

The reconstruction of proto-languages was based on analyses of roots, words, names, phonemes, expressions; structure, syntax, morphology, and grammar of ancient, medieval, and modern languages; as well as on the written evidences and comparative studies of languages, resulting in the following groups:

Afroasiatic languages

Semitic (Abi Ishaq al-Hadrami, 700-730; Sibawayh, 785-795);

Berber (A Basset, 1952; M Quitout, 1997; A El Mountassir, 2003)

Near East languages

Caucasian (SA Starostin and IM Diakonoff, 1986);

Kartvelian (BG Hewitt, 1995; G Kartozia, 2005)

Sino-Tibetan languages

Chinese (Confucius, 485-480BC; Xunzi, 260-240BC; Ma Jianzhong, AD1898);

Tibeto-Burman (RA Miller, 1974; G van Driem, 2003; R Bielmeier and F Haller, 2007)

Eurasian languages

Basque (W von Humboldt, 1810-20);

Dravidian (R Caldwell, 1865; B Krishnamurti, 2003);

Indo-European languages (F Bopp, 1816-56; JLC Grimm, 1819; F de Saussure, 1878; A Meillet, 1897-1925) comprising

(i) Anatolian (HC Melchert, 1994; S Luraghi, 1998; RSP Beekes and MAC de Vaan, 2011),

(ii) Hellenic (Socrates, 405-400BC; Plato, 360-350BC; Aristotle, 335-323BC; Dionysius Thrax, c.100BC; Apollonius Dyskolos, c.AD150),

(iii) Indo-Iranian (Pānini, c.600BC; Kātyāyana, c.250BC; Patañjali the Grammarian, c.140BC; Bhartrihari, c.AD600; W Jones, 1772-87; JF Staal, 1967-2006),

(iv) Italic (A Donatus, c.AD350; Dante Alighieri, 1318-20; J Gilliéron, 1902-21; G Calinescu, 1932-65; S Paliga, 1986-2006),

(v) Celtic (H Wagner, 1958-69; V Kruta, 2000; P Foster and A Toth, 2003; M Albu and I Gauntlett, 2006-10),

(vi) Germanic (the Venerable Bede, 731; JLC Grimm and WC Grimm, 1812-54; IA Aasen, 1836-85; K Verner, 1877; AN Chomsky, 1960-87),

(vii) Armenian (HJ Hübschmann, J Dum-Tragut, 2009),

(viii) Tocharian (D Ringe, 1996; S Nordhoff *et al.*, 2013),

(ix) Balto-Slavic (A Schleicher, 1853-68),

(x) Albanian (Paulus Angelus, 1462; N Iorga, 1915; J Kolgjini, 2004)

North Asian languages

Turkic (Abu al-Ghazi Bahadur, 1778; GJ Ramstedt, 1952-57; L Johanson, 1998);

Mongolic (N Poppe, 1964; J Janhunen, 2003);

Uralic (K Rédei, 1999; A Marcantonio, 2002; H Tommola, 2010);

Chukotko-Kamchatkan (MD Fortescue, 2005-11)

American languages

Eskimo-Aleut (KJV Rasmussen, 1921-24; JH Holst, 2005);

Algonquian (L Bloomfield, 1946; D Pentland, 2006);

Iroquoian (F Lounsbury, 1978; M Mithun, 1985);

Uto-Aztecan (LR Cambell, 1997);

Mayan (LR Campbell and J Justeson, 1984);

Oto-Manguean (DG Brinton, 1871-86; LR Campbell and V Grondona, 2012)

Pacific Rim languages

Pama-Nyungan (N Evans and P McConvell, 1999; N Evans, 2003);

Austronesian (R Blust, 1999; M Ross, 2009);

Tai-Kadai (J Edmondson and D Solnit, 1988; A Diller *et al.*, 2008);

Hmong-Mien (J Enwall, 1995; M Ratliff, 2010);

Mon-Khmer (H Shorto *et al.*, 2006; P Sidwell, 2008)

Recent data show that: 1st, *Spoken languages* are: 2034 in Asia, 1995 in Africa, 1341 in Pacific areas, 949 in Americas, and 209 in Europe; 2nd, *Official languages* prioritized according to associated populations given in millions as follows: English 1400, Chinese 1070, Hindi 700, Spanish 280, Russian 270, French 220, Arabic 170, Portuguese 160, Malay 160, Bengali 150, Japanese 120, etc.; 3rd, *World's number of living languages* is 6909; 4th, *Great monolingual dictionaries* are the *Oxford English Dictionary* (3rd edition) with c.750,000 words; Dutch *Woordenboek der Nederlandsche Taal* with hundreds of thousands headwords and over a million quotes from sources; German *Deutsches Wörterbuch* with over 330,000 headwords in 67,000 print columns; and French *Larousse* with 135,000 definitions and 6,000 notes, 92,000 synonyms and 29,000 antonyms; 5th, *Great bilingual dictionaries* are the English-German *Muret-Sanders* published by *Langenscheit* with over 400,000 words and expressions; and Chinese-Japanese *Dai Kan-Wa jiten*, with more than 50,000 Chinese characters and 500,000 compound words.

References

c.600BC: Pānini, Indian grammarian, was author of *Ashtadhyayi* 'Eight Lectures', a grammar of Sanskrit comprising 4,000 aphoristic statements which provide the *rules of word formation* and, to a lesser extent, *sentence structure*; representing the basis of all later Sanskrit grammars.

485-480BC: Confucius, Chinese philosopher, emphasized the moral commitment implicit in a name, *zhengming*, as a contribution to the *Chinese linguistics*.

450-440BC: Herodotus, Greek historian, travelled to Thrace, the Black Sea's coasts, Persia, Tyre, Egypt, Cyrene, Sicily, and southern Italy, collecting precious ethnographical information for his *Histories*.

405-400BC: Socrates, Greek philosopher, worked on *linguistics*, and introduced *dialectics as a new text genre*.

360-350BC: Plato, Greek philosopher, argued that words denote concepts which are eternal and exist in the world of ideas, such as exposed in his *Cratylus dialogue*; introduced the word *etymology* to describe the history of a word's meaning; and used the word *grammar* in its original meaning as Τέχνη Γραμματική 'The Art of Writing'.

335-323BC: Aristotle, Greek philosopher and scientist, supported the conventional origins of meaning, defined the *logic of speech* and the argument, and worked on *rhetoric* and *poetics*.

260-240BC: Xunzi (Master Xun), Chinese Confucian philosopher, wrote elaborately argued essays, including the 'proper use of terms' from *zhengming*, and adopted a conventional view for the *origin of the sound-to-meaning mapping*, although the objects signified by the term remain real.

c.250BC: Kātyāyana, Ancient Indian Sanskrit grammarian and Vedic priest, elaborated the *Varttika*, based on Pānini's grammar, which became a core part of the *Vyākarana* 'Grammar' canon.

c.140BC: Patañjali the Grammarian, Indian linguist, wrote a *substantial commentary* on Pānini's *Ashtadhyayi*, contributing to the grammarian studies.

138-125BC: Zhang Qian, Chinese statesman, travelled through Xiongnu's territory, and then to the north of Bactria and Central Asia, making notes about the local populations there with their different *social organizations*, *behaviours*, *languages* and *cultures*.

c.100BC: Dionysius Thrax, Greek grammarian, produced the *Techne Grammatike* 'Art of Letters', as the basis of all European works on grammar.

AD89-93: Gaius Cornelius Tacitus, Roman statesman and historian, travelled widely through Germany, and wrote the book *De origine et situ Germanorum*, or simply called *Germania*, recording valuable ethnographic matter, laws and customs of various tribes there, outside the Roman Empire.

c.150: <u>Apollonius Dyskolos</u>, Alexandrian grammarian, first reduced *Greek syntax* to a system, and wrote the treatise entitled *On Syntax*, and shorter works on *pronouns, conjunctions,* and *adverbs.*

170-190: <u>Claudius Galenus</u>, Greek physician, performed many *dissections of the nervous system* in a variety of species, including the ape, and wrote *De usu partium* 'The Uses of the Parts' as a hymn to the creator, whereby the organs of the body were seen perfectly adapted to the functions which they served.

c.350: <u>Aelius Donatus</u>, Roman grammarian, produced treatises on Latin grammar entitled *Ars grammatica*, as the only textbook used in schools of the Middle Ages, so that *Donat* in western Europe came to mean a 'grammar book'.

c.600: <u>Bhartrihari</u>, Hindu poet and philosopher, studying the Sanskrit language and its rules, he produced an improved *Sanskrit grammar.*

640-660: <u>Xuan Zang</u> (Hsüan Tsang), Chinese Buddhist traveller, made a 16-year pilgrimage through China and India, and studied the *life styles, behaviours* and *customs* of their inhabitants.

700-730: <u>'Abd Allah ibn Abi Ishaq al-Hadrami</u>, Arabic grammarian, compiled a *prescriptive grammar* by referring to usage of the Bedouins, whose language was seen as especially pure; his work became influential upon later grammarians.

731: <u>Bede the Venerable</u>, Anglo-Saxon scholar, theologian and historian, produced his greatest work entitled *Historia Ecclesiastica Gentis Anglorum* 'Ecclesiastical History of the English People', as the first source for the origin of English language.

785-795: <u>Sibawayh</u> (Uthman ibn Qunbar Al-Bisri), Persian Muslim linguist and grammarian of the Arabic language, made a detailed description of Arabic in his monumental work *al-Kitāb fī an-nahw* 'The Book of Grammar'.

1271-95: <u>Marco Polo</u>, Venetian merchant and traveller, crossed Central Asia and Gobi Desert to China, where served as envoy to Yunnan, northern Burma, Karakorum, Cochin-China, and southern India, and as Governor of Yang Chow helping to subdue the city of Saianfu; then he returned to Venice, and was prisoner at Genoa, where he wrote *Divisament dou Monde* 'Description of the World', an important account of his systematic observations on nature, *anthropology,* and geography of the visited regions, which was presumably recounted to Rustichello da Pisa by Polo when they were both in the Genoese prison, and later published; the readers of this work named him 'the father of modern anthropology'.

1318-20: <u>Dante Alighieri</u>, Italian poet, wrote the unfinished work *De Vulgari Eloquentia* 'On the Eloquence of Vernacular', discussing the origin of language, the divisions of languages, and the dialects of Italian in particular, extending the scope of linguistic enquiry *from Latin/Greek to include the languages* of his time.

1462: <u>Paulus Angelus</u> (Pal Engjëlli), Archbishop of Durrës, authorized *Formula e Pagëzimit* 'Baptismal Formula', the earliest known text in Albanian language.

1499: <u>Johannes Peyligk</u>, German philosopher, wrote *Compendium philosophiae naturalis*, containing eleven woodcuts which depict the *dura mater* and *pia mater*, as well as the *brain-ventricles*.

1502-10: <u>Ludovico de Varthema</u>, Italian traveller and writer, collected social and cultural information of many peoples living in Arabia, Persia, India, and the Spice Islands, such as recorded in his work *Itinerario de Lodovico de Varthema Bolognese* 'Travels of Ludovico de Varthema'.

1543: <u>Andreas Vesalius</u>, Belgian anatomist, produced his greatest work *De Humani Corporis Fabrica* 'On the Structure of the Human Body', including detailed images of the brain-ventricles, *cranial nerves*, pituitary gland, meninges, and an image of the *peripheral nerves*.

1630-31: <u>John Amos Comenius</u>, Czech educationist, pioneered the new *language teaching methods*, and published *Janua Linguarum Reserata*.

1746-54: <u>Étienne Bonnot de Condillac</u>, French philosopher and psychologist, published *Essai sur l'origine des connaissances humaines* 'Essay on the Origin of Human Knowledge', and *Traité des sensations* 'Treatise on Sensations'.

1772-87: <u>William Jones</u>, English jurist and Orientalist, published the *Persian Grammar*, and pointed out striking resemblance of Sanskrit to Latin and Greek languages, becoming a pioneer of *comparative linguistics*.

1773-92: <u>James Burnett Monboddo</u>, Scottish judge and anthropologist, published six-volume *Of the Origin and Progress of Language*, as a learned but idiosyncratic, but his theory of human affinity with monkeys anticipated the theory of evolution, and the *modern science of anthropology*.

1778: <u>Abu al-Ghazi Bahadur</u>, Uzbek Khan of Khiva, made known the work entitled *Genealogical History of Tatars*, Russian Imperial Academy of Sciences.

1785-98: <u>Immanuel Kant</u>, German philosopher, wrote a book entitled *Grundlagen zur Metaphysik der Sitten* 'Groundwork to the Metaphysic of Morals', presenting the famous imperative 'Act only on that maxim which you can at the same time will to become a universal law', and thus reflecting cerebral functions; and then, based on his lectures for students, published the first major treatise of anthropology, namely *Kants Anthropologie aus pragmatischer Sicht* 'Anthropology from a Pragmatic Point of View', revealing not only his unique contribution to the newly emerging discipline, but also his desire to offer a *practical view of the world and of humanity's place in it*.

1810-20: <u>Wilhelm von Humboldt</u>, German politician and philologist, was the first to study the *Basque language* scientifically, worked on the *languages of the East* and the *South Sea Islands*, wrote *Über die Verschiedenheit des menschlichen Sprachbaues und ihren Einfluß auf die geistige Entwickelung des Menschengeschlechts* 'On the Variety of the Structure of Human

Language and its Influence upon the Mental Development of the Human Race', and formulated a number of 3959 rules of *Sanskrit morphology*.

1812-54: Brothers <u>Jacob Ludwig Carl Grimm</u> and <u>Wilhelm Carl Grimm</u>, German writers and linguists, wrote *Kinder- und Hausmärchen* 'Children and House Tales' as a foundation of the science of comparative folklore; *Deutsche Sagen* 'German Sagas'; *Geschichte der deutsche Sprache* 'History of German Language'; and *Deutsche Wörterbuch* 'German Dictionary'.

1816-56: <u>Franz Bopp</u>, German philologist, produced the Indo-European grammar *Über das Conjugationssystem der Sanskrit-sprache* 'On the System of Conjugation in Sanskrit', tracing common origin of grammatical forms of Indo-European languages; and the greatest six-volume *A Comparative Grammar of Sanskrit, Zend, Greek, Latin, Lithuanian, Old Slavonic, Gothic and German*, later including *Old Armenian*.

1819: <u>Jacob Ludwig Carl Grimm</u>, German philologist, wrote *Deutsche Grammatik* 'German Grammar', considered 'first great scientific linguistic work of the world', and formulated the *Grimm's Law*, which states that in the Proto-Indo-European voiceless stops become voiceless fricatives, voiced stops become voiceless stops, and voiced aspirates become voiced stops or fricatives (depending on context).

1819-41: <u>Arthur Schopenhauer</u>, German philosopher, produced the major work *The World as Will and Idea*, including reflections on the theory of knowledge and implications for the philosophy of nature, aesthetics and ethics; he also asserted that *a dream is a short-lasting psychosis*, and *a psychosis is a long dream*.

1836-76: <u>Gustav Theodor Fechner</u>, German physicist, psychologist and philosopher, studied connections between physiology and psychology, and published *Das Büchlein vom Leben nach dem Tode* 'The Little Book of Life After Death', *Elemente der Psychophysik* 'Elements of Psychophysics', and *Vorschule der Ästhetik* 'Introduction to Aesthetics'.

1836-85: <u>Ivar Andreas Aasen</u>, Norwegian philologist, lexicographer and writer, carried out important studies for replacing the Dano-Norwegian *Riksmål* by *Landsmål*, later known as *Nynorsk*, based on western Norwegian dialects; and published *Grammar of the Norwegian Dialects*, and *Dictionary of the Norwegian Dialects*.

1837: <u>Jan Evangelista Purkinje</u>, Czech physiologist, outlined the key features of cell theory, describing *nuclei* and *Purkinje cells* in the cerebellar cortex, and gave the *first description of the neurons*.

1846-88: <u>William Benjamin Carpenter</u>, English biologist, had the idea of 'unconscious cerebration', and published *Principles of Human Physiology*, *The Microscope and its Revelations*, *Principles of Mental Physiology*, as well as *Nature and Man*.

1847-54: <u>George Boole</u>, English mathematician and logician, published *Mathematical Analysis of Logic*, and *Laws of Thought*, using mathematical symbolism to express logical relations.

1851-77: <u>Lewis Henry Morgan</u>, US anthropologist and social theorist, produced works such as *The League of the Ho-de-no-sau-nee or Iroquois*, Sage and Brothers, Rochester; *The American Beaver and his Works*, J.B. Lippincott and Co., Philadelphia; *Systems of Consanguinity and Affinity of the Human Family*, Smithsonian Institute; and *Ancient Society*, Henry Holt and Co., New York.

1853-68: <u>August Schleicher</u>, German philologist, wrote fundamental works for the studies of Indo-European languages, including *Die ersten Spaltungen des indogermanischen Urvolkes*; *Handbuch der litauischen Sprache*, the first scientific compendium of Lithuanian language; *Kurzer Abriss der Geschichte der italienischen Sprachen*, relating to the Italian language; *Compendium der vergleichenden Grammatik der indogermanischen Sprachen*, relating to the German language; and the great *A Compendium of the Comparative Grammar of the Indo-European, Sanskrit, Greek, and Latin Languages*, attempting to reconstruct the Proto-Indo-European language.

1861-78: <u>Paul Broca</u>, French surgeon and anthropologist, first located the motor speech centre in the brain, since known as the *convolution of Broca* 'Broca's gyrus'; proposed the *motor aphasia*, the meaning of aphasia being the loss of speech marked by severe defect in the understanding of speech; his anthropological investigations gave strong support to CR Darwin's theory of the evolutionary descent of man.

1863-1920: <u>Wilhelm Max Wundt</u>, German physiologist and psychologist, studied the nervous system and senses, relations between physiology and psychology, logic and other subjects; and published *Vorlesungen über die Menschen und Thierseele* 'Lectures on the Mind of Humans and Animals', *Grundriss der Psychologie* 'Outlines of Psychology', and the ten-volume *Völkerpsychologie* 'Ethnic Psychology'; founding *experimental psychology*.

1865: <u>Robert Caldwell</u>, Scottish-Irish Evangelist missionary and linguist, published his work entitled *A comparative grammar of the Dravidian, or, South-Indian family of languages*, Harrison: London.

1868: <u>Louis Lartet</u>, French geologist, discovered the first five skeletons of *Cro-Magnon man* in a rock shelter named Abri de Crô-Magnon within French-Italian Occitania 'the Oc Country', considering that those skeletons belonged to a human race distinct from the Neanderthals.

1870-90: <u>Jean Martin Charcot</u>, French pathologist and neurologist, made important observations on *multiple sclerosis*, *amyotrophic sclerosis* and *muscular atrophy*, being considered one of the fathers of *neurology*.

1871-86: <u>Daniel Garrison Brinton</u>, US linguist, studied languages of Native people of America, and wrote the book *The Arawack Language of Guiana in Its Linguistic and Ethnological Relations*, McCalla and Stavely, Printers, Philadelphia; and the article *Notes on the Mangue; An Extinct Dialect Formerly Spoken in Nicaragua*, Proceedings of the American Philosophical Society, 23(122).

1874-85: Carl Wernicke, German neurologist and psychiatrist, published *Der Aphasische Symtomencomplex* 'The Aphasic Syndrome', revealing *aphasia* 'loss of speech' as marked by a severe defect in understanding of speech, known as *sensory aphasia*, and showing that this kind of aphasia is typically localized in the so-called *Wernicke's area* of the left temporal lobe.

1875: Heinrich Johann Hübschmann, German scholar of Iranian and Armenian studies, published his work *Über die Stellung des armenischen im Kreise der indogermanischen Sprachen*, Zeitschrift für Vergleichende Sprachforschung, 23.

1877: Karl Verner, Danish linguist, formulated the *Verner's Law*, according to which Proto-Germanic fricatives become voiced if the next conditions are met: they are not initials, but preceded and followed by voiced, and accent is not on the immediately preceding syllable.

1878: Ferdinand de Saussure, Swiss linguist, published *Mémoire sur le système primitif des voyelles dans les langues indo-européenes* 'Memoir on the Primitive Vowels in the Indo-European Languages', constituting the first serious attempt to determine the nature of language as the object of which linguistics is the study, and showing the importance of *synchronic analysis*, although this focus has shifted and the term 'philology' has been later used for the study of a language's grammar, history, and literary tradition.

1879: William Macewen, Scottish neurosurgeon, successfully removed a tumour involving the *meninges of the brain*, and published work on the treatment of aneurysms (dilatation of an artery) by acupuncture.

1891-1931: George Frederick Stout, English philosopher and psychologist, edited the journal Mind, and made important contributions to the psychology and philosophy of mind by publications such as *Analytic Psychology*, *Manual of Psychology*, and *Mind and Matter*.

1893-1905: Lazarus Ludwig Zamenhof, Polish oculist and philologist, invented the *Esperanto* 'One who hopes' as an international language to promote world peace, and published *Fundamento de Esperanto* 'Basis of Esperanto'.

1897-1925: Antoine Meillet, French philologist, produced standard works on *Old Slavonic, Greek, Armenian, Old Persian*, etc., as well as on *comparative Indo-European grammar*, and *linguistic theory*; such as *Research on the Use of the Genitive-Accusative in Old-Slavonic*; *Esquisse d'une grammaire comparée de l'arménien classique*; *Caractères généraux des langues germaniques*; and *La méthode comparative en linguistique historique* 'The comparative method in historical linguistics'.

1898: Ma Jianzhong, late Qing Dynasty Chinese official and scholar, was author of *Mashi Wentong* 'Basic principles for writing clearly and coherently by Master Ma', a textbook of *Chinese grammar*, in modern sense, based on the Latin 'prescriptive' model.

1900-23: Sigmund Freud, Austrian neurologist, founded *psychoanalysis*; developed a theory that the content of dreams is driven by unconscious fulfilment of wish; and divided the unconscious mind into *Id, Ego*, and *Super-Ego*.

1902-21: Jules Gilliéron, Swiss linguist, published *Atlas linguistiques de la France; La Généalogie des mots qui désignent l'abeille* 'The Etymology of Words Relating to the Bee'; and *Pathologie et thérapeutique verbales* 'The Pathology and Treatment of Words'.

1909: Harvey Williams Cushing, US neurosurgeon, discovered a new operative approach to the *pituitary gland*, and made a detailed study of its activity, characterizing the effects of *underactivity* that causes dwarfism in a child, and of *overactivity* causing a form of gigantism in adults.

1911-40: Franz Uri Boas, German-US anthropologist, was considered 'the father of modern cultural anthropology'; and wrote *The Mind of Primitive Man; Handbook of American Indian languages*, vol.1, Bureau of American Ethnology, 40; *Folk-tales of Salishan and Sahaptin tribes*, Washington State Library's Classics in Washington History collection; and *Race, Language, and Culture*; outlining new and less simple concepts of *culture* and of *race*, as well as arguing *against racial ideology*.

1913-20: Wolfgang Köhler, German psychologist, was director of the anthropoid research station in the Canary Islands, becoming an authority on *problem-solving in animals*.

1914-33: Leonard Bloomfield, US linguist, played a major part in making *linguistics an independent scientific discipline*, by publishing *Introduction to the Study of Language*, and the major work *Language*, treating on linguistic theory and pioneering the *structuralism*.

1915: Nicolae Iorga, Romanian historian and linguist, discovered and published a standard *Albanian text* entitled *Unte paghesont premenit Atit et Birit et Spertit Senit* 'I baptize you in the name of the Father and Son and the Holy Spirit'.

1919-30: Charles Scott Sherrington and Edgar Douglas Adrian, English physiologists, examined *sensory systems* in animals and humans, investigated the response *mechanisms of receptors and sense organs*, and studied the recording and analysing of *information in the central nervous system*.

1920-40: António Caetano Egas Moniz, Portuguese neurosurgeon, worked on the use of dyes in the X-ray localization of *brain tumours*, and developed *prefrontal lobotomy* for the control of schizophrenia and other mental disorders.

1921-24: Knud Johan Victor Rasmussen, Danish explorer and ethnologist, in support of the theory that *Inuits and North American Indians* are both descendants of *Asian migratory tribes*, he made a cross-examination from Greenland to Bering Strait, by dog sledge, to visit Inuit groups along the route.

1921-30: Joseph Erlanger and Herbert Spencer Gasser, US physiologists, using powerful new electrical equipment to dissect and analyse nature and function of nerve fibres, discovered the *fundamental properties of neural conduction of impulses,* and deduced that the *speed of impulse is proportional to diameter of nerve fibre.*

1924-34: <u>Marcel Maus</u>, French anthropologist and sociologist, studied religion and ancient languages; became a great proponent of 'social ethnology'; produced works such as *Essay sur le don,* and *Les Techniques du corps,* Journal de Psychologie; and initiated theories regarding gift exchange among groups throughout the world, his work 'Essay on the Gift' describing the intrinsic bond forged between giver and receiver.

1926-37: <u>Zora Neale Hurston</u>, US folklorist, anthropologist and novelist, became known for the short stories *Sweat,* and *The Gilded Six-Bits*; and the novel *Their Eyes Were Watching God.*

1928-35: <u>Henry Hallett Dale</u>, English physiologist, and <u>Otto Loewi</u>, German pharmacologist, discovered *acetylcholine*; and identified several other possible *transmitter substances*, respectively, in relation to chemical transmission of nerve impulses.

1928-70: <u>Margaret Mead</u>, US cultural anthropologist, published her works including *Coming of Age in Samoa*; *Anthropology and the Abnormal*, Journal of General Psychology, 10; *Growing Up In New Guinea*; *Male and Female*; and *Culture and Commitment.*

1932-65: <u>George Calinescu</u>, Romanian novelist, linguist, literary critic and historian, produced outstanding works on aesthetics and world literature including *Principles of Aesthetics*, and *The Meaning of Classicism*; as well as his masterpiece *The History of Romanian Literature from its Origins to the Present.*

1934-46: <u>Ruth Fulton Benedict</u>, US anthropologist and folklorist, wrote *Patterns of Culture*, Houghton Mifflin Harcourt; *Race: Science and Politics*; and *The Chrysanthemum and the Sword: Patterns of Japanese Culture*, Rutland, VT and Tokyo

1939-43: <u>Louis Trolle Hjelmslev</u>, Danish linguist, devised a system of linguistic analysis known as *glossematics*, which was based on the study of the distribution of, and the relationships between, the smallest meaningful units of a language, called *glossemes*; and wrote *Prolegomena to a Theory of Language.*

1940-50: <u>Ulf Svante von Euler</u>, Swedish pharmacologist, studied the *neurally active chemicals*; isolated and characterized the principal transmitter of the sympathetic nervous system, called *noradrenaline*; and studied its properties and chemical relatives.

1943-60: <u>Kenneth Lee Pike</u>, US linguist, developed a system of linguistic analysis known as *tagmemics*, and wrote the books *Phonetics, Phonemics, Tone Languages*, and *Language in Relation to a Unified Theory of the Structure of Human Behaviour.*

1945-49: <u>Walter Rudolf Hess</u>, Swiss physiologist, and <u>António Caetano Egas Moniz</u>, Portuguese neurologist, established methods of stimulating localized areas of brain by means of fine needle electrodes to study *brain function*; and used dyes in X-ray localization of brain tumours and development of *prefrontal lobotomy* for control of schizophrenia and other mental diseases.

1945-55: Jean Malaurie, French explorer and anthropogeographer, led over 30 scientific expeditions to the Arctic regions, made several *documentary films on the Inuit people*, and became famous for his book *The Last Kings of Thule*.

1945-70: Bernard Katz, German-born British biophysicist, worked on mechanisms of neural transmission, showing that *chemical neurotransmitters are stored in nerve terminals* and *released in specific portions* called 'quanta' *when stimulated by the arrival of a neural impulse*.

1946: Leonard Bloomfield, US linguist, studied Native American languages, and wrote *Algonquian*, in 'Linguistic Structures of Native America', Harry Hoijer, ed., Viking Fund Publications in Anthropology, 6, New York.

1948-50: Alan Lloyd Hodgkin and Andrew Fielding Huxley, English physiologists, described in physical-chemical and mathematical terms the mechanisms by which *nerves conduct electrical impulses* by the movement of electrically charged particles across the nerve membrane, opening the way to study and understand different kinds of excitable membranes.

1951-63: John Carew Eccles, Australian neurophysiologist, recorded *depolarization of a post-synaptic muscle fibre* in response to a neural stimulus, identified *inhibitory neurons*, and demonstrated how *inhibitory synapses control the flow of information* within the nervous system.

1951-89: Herbert Marshal McLuhan, Canadian philosopher of *communication theory*, became known for coining expressions such as 'the medium is the massage' and 'the global village', and for predicting the World Wide Web; his published works include *The Mechanical Bride: Folklore of Industrial Man*; *The Gutenberg Galaxy: The Making of Topographic Man*; *Understanding Media: The Extensions of Man*; *The Medium is the Massage: An Inventory of Effects*; *War and Peace in the Global Village*; *From Cliché to Archetype*; and *The Global Village: Transformations in World Life and Media in the 21st Century*.

1952: André Basset, French linguist, wrote his work *La langue berbère*, published in 'Handbook of African Languages', Part I, Daryll Forde ed., London: Oxford University Press.

1952-57: Gustaf John Ramstedt, Swedish-Finnish linguist and diplomat, published his two-volume *Einführung in die altaische Sprachwissenschaft* 'Introduction to Altaic Languages', Helsinki: Suomalais-Ugrilainen Seura.

1953-90: Hans Jürgen Eysenck, British psychologist, published *Uses and Abuses of Psychology*, *Know Your Own IQ*; *Race, Intelligence and Education* - a controversial view on racial differences in intelligence; and autobiographical *Rebel with a Cause*.

1955-65: Roger Wolcott Sperry, US neuroscientist, pioneered behavioural investigation of 'split-brain' animals and humans, established that each hemisphere possesses specific higher functions: the *left side* controlling verbal activity and processes such as writing and reasoning; whereas the *right side* is more responsive to music, face and voice recognition.

1958-69: <u>Heinrich Wagner</u>, Swiss-born Irish linguist, published his monumental volume *Linguistic Atlas and Survey of Irish Dialects*; studies *Gaeilge Theilinn*, and *Das Verbum in den Sprachen der Britischen Inseln*; as well as collections of proverbs *Sean-Chaint Theilinn*, and *Sean-chaint na gCruach*.

1958-78: <u>Claude Lévi-Strauss</u>, French anthropologist and ethnologist, was the 'founder of structuralism', and published his works *Anthropologie structurale* 'Structural Anthropology', *Mythologiques I - IV*, *Anthropologie structurale deux*, vol. II, and *Myth and Meaning*, Routledge & Kegan Paul, UK, Taylor & Francis Group.

1960-83: <u>Clifford James Geertz</u>, US anthropologist, investigated symbolic, or interpretative, anthropology; and wrote *Religion of Java*, University of Chicago Press; *Agricultural Involution: the process of ecological change in Indonesia*, *The Interpretation of Cultures*, Basic Books; *Kinship in Bali*; and *Local Knowledge*.

1960-87: <u>Noam Chomsky</u>, US linguist, contributed at linguistic theory by works such as *Aspects of the Theory of Syntax*; *Cartesian Linguistics*; *The Sound Pattern of English*; *Language and Mind*; *Reflections on Language*; *The Logical Structure of Linguistic Theory*; and *Language and Problems of Knowledge*; by which he established the *generative theory* of linguistics.

1964: <u>Nicholas Poppe</u>, Russian linguist, published the textbook *Grammar of Written Mongolian*, Wiesbaden: Harrassowitz.

1967-2006: <u>Johan Frederik Staal</u>, Dutch-born US scholar of Greek and Indian logic and philosophy, and Sanskrit grammar, studied the methods of linguistic theory, and the applications of modern mathematical logic to linguistics, stating that 'Pāṇini is the Indian Euclid'; and published *Word Order in Sanskrit and Universal Grammar*; *A Reader on the Sanskrit Grammarians*; *Universals - Studies in Indian Logic and Linguistics*; and *Artificial Languages across Sciences and Civilizations*.

1969: <u>Richard Erskine Frere Leakey</u>, Kenyan paleoanthropologist, discovered remains of the oldest modern humans, which were first presented within his article *Early Homo sapiens remains from the Omo River region of south-west Ethiopia*, Nature, 222.

1969-82: <u>Eric Robert Wolf</u>, Austrian anthropologist, wrote *Peasant Wars of the Twentieth Century*, Harper & Row; and *Europe and the People without History*, University of California Press.

1974: <u>Roy Andrew Miller</u>, US linguist, published his article *Sino-Tibetan: Inspection of a Conspectus*, Journal of the American Oriental Society, 94(2).

1978: <u>Floyd Lounsbury</u>, US linguist and anthropologist, wrote *Iroquoian Languages*, in Bruce Trigger 'Handbook of North American Indians, vol. 15: Northeast', Washington DC: Smithsonian Institution.

1979: <u>Mary Ainsworth</u>, US personality psychologist, made known some results of her research on social competence, by a work entitled *Infant-mother attachment*, American Psychologist, 34.

1981-92: <u>Alfred Goodman Gilman</u>, US pharmacologist, and <u>Martin Rodbell</u>, US biochemist, discovered the *G-proteins* that enable cells to respond to signals from other cells or from outside the body.

1984: <u>Lyle Richard Campbell</u> and <u>John Justeson</u>, US linguists, edited their work *Phoneticism in Mayan Hieroglyphic Writing*, Institute for Mesoamerican Studies, 9, SUNY Albany/University of Texas Press.

1985: <u>Tetsuro Matsuzawa</u>, Japanese linguist and cognitive scientist, wrote the article entitled *Colour naming and classification in a chimpanzee (Pan Troglodytes)*, Journal of Human Evolution, 14(3). \ <u>Marianne Mithun</u>, US linguist, wrote the article *Untangling the Huron and the Iroquois*, International Journal of American Linguistics, 51(4).

1986: <u>Serghei Anatolyevich Starostin</u> and <u>Igor Mikhailovich Diakonoff</u>, Russian historical linguists, wrote the book *Hurro-Urartian as an Eastern Caucasian Language*, Munich: R Kitzinger.

1986-2006: <u>Sorin Paliga</u>, Romanian linguist, published works including *Etymological Lexicon of the Indigenous (Thracian) Elements in Romanian*, and *Roman and Pre-Roman Influences in the Southern Slavonic Languages*.

1988: <u>Jerold Edmondson</u> and <u>David Solnit</u>, US linguists, edited *Comparative Kadai: linguistic studies beyond Tai branch*, Dallas: Summer Institute of Linguistics, 86.

1991: <u>Diana Baumrind</u>, US psychologist, published her article *The influence of parenting style on adolescent competence and substance use*, Journal of Early Adolescence, 11(1).

1994: <u>Harold Craig Melchert</u>, US linguist, produced the work entitled *Anatolian historical phonology*, Leiden studies in Indo-European, 3, Amsterdam: Rodopi.

1995: <u>Christopher Williams</u>, <u>Michael Revow</u>, and <u>Geoffrey Hinton</u>, Canadian neurologists, researched neural networks, and published their results under the title *Using a neural net to instantiate a deformable model*, Advances in Neural Information Processing Systems, 7. \ <u>Brian George Hewitt</u>, English linguist and Professor of Caucasian languages, wrote a treatise entitled *Georgian: A Structural Reference Grammar*, Amsterdam and Philadelphia: John Benjamins Publishing Company. \ <u>Joakim Enwall</u>, Swedish linguist, published his work *Hmong writing systems in Vietnam: a case study of Vietnam's minority language policy*, Stockholm, Sweden: Center for Pacific Asian Studies.

1996: <u>Robin Ian MacDonald Dunbar</u>, British anthropologist and psychologist, wrote the book *Grooming, Gossip and the Evolution of Language*, London: Faber and Faber. \ <u>Hans Henrich Hock</u> and <u>Brian Joseph</u>, US linguists, produced an outstanding work entitled *Language History, Language Change, and Language Relationship: An introduction to Historical and Comparative Linguistics*, as a comprehensive study of languages. \ <u>Donald Ringe</u>, US linguist and Indo-Europenist, studied the extinct *Tocharian language*, and wrote *On the Chronology of Sound Changes in Tocharian, vol.I: From Proto-Indo-European to Proto-Tocharian*, New Haven, CT: American Oriental Society.

1996-2000: Mario Alinei, Italian linguist, published two-volume *Origini delle Lingue d'Europa* 'The Origins of Language of Europe', formulating *Palaeolithic Continuity Theory* with proposal that Indo-Europeans arrived in Europe tens of thousands of years ago, and that by the end of Ice Age already differentiated into local language speakers occupying territories within or close to their now-traditional homelands; and hypothesizing that a homogeneous early Indo-European people appeared in Europe about 6000 years ago; he was founder and editor of *Quaderni di semantica*, a journal of theoretical and applied semantics.

1997: Michel Quitout, French linguist, published his book *Grammaire berbère (rifain, tamazight, chleuh, kabyle)*, Paris-Montréal: Éditions l'Harmattan. \ Lyle Richard Campbell, US linguist, studied languages of Native people of America, and wrote the book *American Indian Languages: The Historical Linguistics of Native America*, Oxford University Press.

1998: Ib Ulbæk, Danish linguist, as an adept of 'continuity theories', he wrote a chapter entitled *The Origin of language and cognition*, which was published in JR Hurford, M Studdert-Kennedy and C Knight 'Approaches to the Evolution of Language: Social and Cognitive Bases', Cambridge University Press. \ Andrew Dalby, English linguist, published his book *Dictionary of Languages, The Definitive Reference to more than 400 Languages*, Bloomsbury Publishing plc, London. \ Silvia Luraghi, Italian linguist, published her work *The Anatolian Languages*, in Anna Giacalone Ramat and Paolo Ramat eds. 'The Indo-European Languages', Routledge Language Family Descriptions, London and New York: Routledge. \ Lars Johanson, Swedish linguist, wrote *The History of Turkic*, in Lars Johanson and Éva Ágnes Csató, eds., 'The Turkic Languages', London-New York: Routledge.

1999: Károly Rédei, Hungarian linguist, published his work *Zu den uralisch-jukagirischen Sprachkontakten*, Finnisch-Ugrische Forschungen, 55. \ Nicholas Evans and Patrick McConvell, Australian linguists, wrote *The Enigma of Pama-Nyungan Expansion in Australia*, Archaeology and language, vol. 29, Roger Blench and Matthew Spriggs, eds., Routledge. \ Robert Blust, US historical linguist and ethnologist, published his work *Subgrouping, circularity and extinction: some issues in Austronesian comparative*, in E Zeitoun and PJK Li 'Selected papers from the Eighth International Conference on Austronesian Linguistics', Taipei: Academia Sinica. \ Robert Crag, US communication theorist, extended the *communication theory*, as presented in his work *Communication Theory as a Field*, International Communication Association.

1999-2003: Paul Edward Farmer, US anthropologist and physician, wrote *Infections and Inequalities: The Modern Plagues*, and *Pathologies of Power: Health, Human Rights, and the New War on the Poor*, both published at

133

Berkeley: University of California Press.

2000: <u>Venceslas Kruta</u>, French linguist, studied changes in Celtic languages and published his work *Les Celtes – Histoire et dictionnaire, Des origines à la romanisation et au christianisme*, an important book for Celtic culture and language. \ <u>Glanville Price</u>, professor of French at the University of Wales Aberystwyth, edited *Encyclopedia of the Languages of Europe*, Blackwell Publishers Ltd.

2001: <u>Christopher Boehm</u>, US anthropologist, wrote a series of books related to human behaviour, including *Hierarchy in the Forest. The Evolution of egalitarian behaviour*, Cambridge, MA: Harvard University Press. \ <u>Peter Dayan</u>, US physiologist and cellular biophysicist, and <u>Larry Abbott</u>, US neuroscientist, published *Theoretical Neuroscience: Computational and mathematical Modelling of Neural Systems*, MIT Press. \ <u>Cecilia Lai</u>, <u>Simon Fisher</u>, <u>Jane Hurst</u>, <u>Faraneh Vargha-Khadem</u> and <u>Anthony Monaco</u>, English psychologists and geneticists, discovered the FOXP2 gene responsible for speech and language, and wrote the article *A forkhead-domain gene is mutated in a severe speech and language disorder*, Nature, 413.

2001-04: <u>Andrei Alexandrescu</u>, Romanian C++ programmer and author, working on *policy-based design* implemented via *template metaprogamming*, developed an initial *computer programming language (C)* into the *C plus plus (C++) programming languages*; and wrote the books *Modern C++ Design: Generic Programming and Design Patterns Applied*; and *C++ Coding Standards: 101 Rules, Guidelines, and Best Practices*, both published by Addison-Wesley.

2002: <u>Stephen Anderson</u> and <u>David Lightfoot</u>, US linguists, published their book *The Language Organ: Linguistics as Cognitive Physiology*, Cambridge University Press. \ <u>Marc Hauser</u>, <u>Avram Noam Chomsky</u> and <u>Tecumseh Fitch</u>, US psychologists and linguists, studied the high-level reference systems, wrote the article *The Faculty of Language: What Is It, Who Has It, and How Did It Evolve?*, Science, 298. \ <u>Derek Bickerton</u>, English linguist, published his work *Foraging Versus Social Intelligence in the Evolution of Protolanguage*, in Alison Wray, ed., 'The Transition to Language', Oxford University Press. \ <u>Angela Marcantonio</u>, Italian historical linguist, wrote the article *The Uralic Language Family: Facts, Myths and Statistics*, Transactions of the Philological Society, 35, Oxford and Boston: Blackwell.

2003: <u>Tim White</u>, US; <u>Berhane Asfaw</u>, Ethiopian; <u>David DeGusta</u>, <u>Henry Gilbert</u> and <u>Gary Richards</u>, US; <u>Gen Suwa</u>, Japanese; and <u>Clark Howel</u>, US, anthropologists and palaeontologists, after six years of research on the human remains discovered at Herto Bouri, they published their work *Pleistocene Homo sapiens from Middle Awash, Ethiopia*, Nature, 423. \ <u>Abdallah El Mountassir</u>, Moroccan linguist, studied the *Berber language*, and published a book entitled *Dictionnaire des verbes Tachelhit-Français (parler berbère du sud du Maroc)*, l'Harmattan, Paris-Budapest-Torino. \ <u>George van Driem</u>, Swiss linguist, wrote *Tibeto-Burman Phylogeny and Prehistory: Languages, Material Culture and Genes*, in Peter Bellwood and Colin Renfrew, eds.,

'Examining the farming/language dispersal hypothesis'. \ Juha Janhunen, Finnish linguist, was editor of the book *The Mongolic Language*, Routledge Language Family Series, London; Routledge. \ Bhadriraju Krishnamurti, Indian linguist and Dravidianist, published his book *The Dravidian Languages*, Cambridge Language Surveys, 1st ed., Cambridge University Press. \ Nicholas Evans, Australian linguist, edited *The Non-Pama-Nyungan Languages of Northern Australia. Comparative studies of the continent's most linguistically complex region*, Canberra: Pacific Linguistics. \ Peter Foster and Alfred Toth, US linguists, published their work entitled *Toward a phylogenetic chronology of ancient Gaulish, Celtic, and Indo-European*, Proceedings of the National Academy of Sciences of the USA, 100(15).

2004: Peter Brown, Australian; Thomas Sutikna, Indonesian; Michael Morwood, Australian; Radien Soejono, Jatmiko, Wahyu Saptomo and Rokus Awe Due, Indonesians, anthropologists and palaeontologists, discovered and described human remains of *Homo floresiensis* 'Hobbit', as published in their article *A new small-bodied hominin from the Late Pleistocene of Flores, Indonesia*, Nature, 431. \ Avram Noam Chomsky, US theoretical linguist, as a proponent of 'discontinuity theories', published his work *Language and Mind: Current thoughts on ancient problems*, I and II, in L Jenkins (ed.) 'Variation and Universals in Biolinguistics', Amsterdam: Elsevier. \ Hansell Stedman, Benjamin Wesley Kozyak, Amy Nelson, Danielle Thesier, Leonard Su, David Low, Charles Bridges, Joseph Shrager, Nancy Minugh-Purvis and Mark Mitchell, geneticists, paediatrists and anthropologists in the Department of Surgery, University of Pennsylvania, Philadelphia, USA, wrote an article entitled *Myosin gene mutation correlates with anatomical changes in the human lineage*, Nature, 428. \ Alessandro Duranti, US Professor of anthropology, edited the work entitled *Companion to Linguistic Anthropology*, Malden, MA: Blackwell. \ Julie Kolgjini, US linguist, presented the thesis entitled *Palatalization in Albanian: An Acoustic Investigation of Stops and Affricates*, Dissertation, University of Texas at Arlington.

2005: Tao Gong and William Wang, Hong Kong linguists, wrote their article entitled *Computational Modelling on Language Emergence: A Co-evolution Model of Lexicon, Syntax and Social Structure*, Language and Linguistics, 6(1). \ Guram Kartozia, Georgian linguist, published his work *The Laz language and its place in the system of Kartvelian languages*, Tbilisi. \ Jan Henrik Holst, German linguist, studied the Eskimo-Aleut languages, and wrote *Einführung in die eskimo-aleutischen Sprachen*, Hamburg: Buske.

2005-11: Michael David Fortescue, British-born Danish linguist, produced two important works on Chukotko-Kamchatkan languages, one *Comparative Chukotko-Kamchatkan Dictionary*, Trends in Linguistics, 23, Berlin: Mouton de Gruyter; and another *The relationship of Nivkh to Chukotko-Kamchatkan revisited*, Lingua, 121(8).

2006: David Pentland, US linguist, published the part called *Algonquian and Ritwan Languages*, in Keith Brown, ed., 'Encyclopaedia of Languages and Linguistics', 2nd ed., Amsterdam: Elsevier. \ Harry Shorto, British, Paul

Sidwell, Australian, <u>Doug Cooper</u>, US, and <u>Christian Bauer</u>, Thai, linguists, edited *A Mon-Khmer Comparative Dictionary*, Canberra: Australian National University, Pacific Linguistics.

2006-10: <u>Marius Albu</u>, Romanian-born British researcher of old Celtic language, published his work *Analogies between Celtic and Eastern Languages*, University Book ed.; and together with his daughter <u>Ioana Gauntlett</u> unveiled a significant number of still surviving Celtic place names, words and roots in many Eastern European and even in Turkish, which were presented in their book *Celtic Names in Western and Eastern European Languages - Evidences for Cultural Diffusion*, The Edwin Mellen Press, Lewiston-Queenston-Lampeter.

2007: <u>Chris Knight</u> and <u>Camilla Power</u>, English anthropologists, published their work *Social Conditions for the Evolutionary Emergence of Language*, The Oxford Handbook of Language Evolution, Chapter 37. \ <u>Margaret Semrud-Clikeman</u>, US child neuropsychologist, published a book entitled *Social competence in children*, New York: Springer. \ <u>Roland Bielmeier</u> and <u>Felix Haller</u>, Swiss linguists, edited *Linguistics of the Himalayas and Beyond*, Berlin and New York: Mouton de Gryter.

2008: <u>John Fleagle</u>, from Stony Brook University, NY, <u>Zelalem Assefa</u>, from the National Museum of Natural History, Washington D.C., <u>Francis Brown</u>, from the College of Mine and Earth Sciences, University of Utah, and <u>John Shea</u>, from the Department of Anthropology, Stony Brook University, NY, USA, made stratigraphic, palaeontological, anthropological and radiometric researches on the fossil *Omo* remains, and wrote the paper entitled *Palaeoanthropology of the Kibish Formation, southern Ethiopia: Introduction*, Journal of Human Evolution, 55(3). \ <u>Anthony Diller</u>, Australian, <u>Jerold Edmondson</u>, US, and <u>Yongxian Luo</u>, Australian, linguists, edited *The Tai-Kadai Languages*, London-New York, Routledge. \ <u>Paul Sidwell</u>, Australian linguist, was author of *Issues in the morphological reconstruction of Proto-Mon-Khmer*, in Claire Bowern *et al.*, eds., 'Morphology and language history: in honour of Harold Koch', Philadelphia, John Benjamins Publishing Company. \ <u>Stephen Littlejohn</u> and <u>Karen Foss</u>, US communication theorists, published their book *Theories of Human Communication*, 9th edition, Thomson and Wadsworth.

2009: <u>Dan Jurafsky</u> and <u>James Martin</u>, US computational linguists, published their work *Speech and language processing: An introduction to natural language processing, computational linguistics, and speech recognition*, 2nd edition, Upper Saddle River, New Jersey: Pearson Prentice Hall. \ <u>Derek Bickerton</u>, English linguist, studied the Creole language, and defined the *protolanguage* as presented in his work *Adam's Tongue: How Humans Made Language, How Language Made Humans*, New York: Hill and Wang/Farrar, Straus and Giroux. \ <u>Jasmine Dum-Tragut</u>, Austrian linguist, wrote the book *Armenian: Modern Eastern Armenian*, Amsterdam: John Benjamins Publishing Company. \ <u>Malcolm Ross</u>, British-Australian linguist, published his work *Proto Austronesian verbal morphology: A reappraisal*, in Alexander

Adelaar and Andrew Pawley, eds., 'Austronesian historical linguistics and culture history: a festschrift for Robert Blust', Canberra Pacific Linguistics.

2010: Johannes Krause and Qiaomei Fu, from Germany; Jeffrey Good, from the USA; Bence Viola, from Austria; Michael Shunkov and Anatoli Derevianko, from Russia; and Svante Pääbo, from Germany, anthropogeneticists, made genomic analyses of a finger-bone from the *Denisova hominin*, found by Russian archaeologists in 2008 and carbon-dated 41000 years ago, which revealed a hitherto-unknown migration out of Africa about one million years ago; and published their results within a paper entitled *The complete mitochondrial DNA genome of an unknown hominin from southern Siberia*, Nature, 464. \ Wu Liu, Chang-Zhu Jin, Ying-Qi Zhang, Yan-Jun Cai, Song Xing, Xiu-Jie Wu and Hai Cheng, Chinese; Lawrence Edwards, US; Wen-Shi Pan, Da-Gong Quin and Zhi-Sheng An, Chinese; Erik Trinkaus, US; and Xin-Zhi Wu, Chinese, archaeologists and anthropologists, researched the modern human presence in China, and wrote the article *Human remains from Zhirendong, South China, and modern human emergence in East Asia*, Proceedings of the National Academy of Sciences of the USA, 107(45). \ Michael Arbib, US computer scientist, and Derek Bickerton, English linguists, edited the work *The Emergence of Protolanguage: Holophrasis vs Compositionality*, John Benjamins Publishing Company. \ Hannu Tommola, Finnish linguist, published his work *Finnish among the Finno-Ugrian languages*, in 'Mood in the Languages of Europe', John Benjamins Publishing Company. \ Martha Ratliff, US linguist, wrote a study entitled *Hmong-Mien-language history*, Canberra, Australia: Pacific Linguistics.

2011: François Mairesse and Marilyn Walker, US researchers on natural linguistic processing and machine learning scientists, produced the work *Controlling User Perceptions of Linguistic Style: Trainable Generation of Personality Traits*, Computational Linguistics, 37(3). \ Andrew Lowler, English archaeologist, studied the possible routes of migration of *Homo sapiens* out of Africa, resulting in his published work *Did Modern Humans Travel Out of Africa via Arabia?*, Science, 28(331). \ Roy Hoshi Hamilton, US physician and neurologist, wrote *Looking at things in a different perspective created the idea of ethics of neural enhancement using noninvasive brain simulation*, Neurology, 76(2). \ Robert Stephen Paul Beekes and Michiel Arnoud Cor de Vaan, Dutch linguists, wrote the book *Comparative Indo-European linguistics: an introduction*, 2nd ed., Amsterdam and Philadelphia: John Benjamins Publishing Company. \ Claude Elwood Shannon, US chemist published his work *A Mathematical Theory of Communication*, The Bell System Technical Journal.

2012: Lyle Richard Campbell and Verónica Grondona, US linguists edited a book entitled *The Indigenous Languages of South America: A Comprehensive Guide*, Berlin: Mouton de Gruyter.

2013: Michael Hammer, associate professor in the University of Arizona, investigated the "Y-chromosomal Adam" of an African American man, and made public his findings under the title *Don't call him 'Adam': South Carolina man's genes help date first man*, FoxNews.com. \ James Steele, Margaret Clegg and Sandra Martelli, English palaeontologists and linguists, examined the morphology of the hyoid bone in three closely related species - *Homo sapiens, Pan troglodytes* and *Gorilla gorilla*, and wrote a book entitled *Comparative morphology of the hominin and African ape hyoid bone, a possible marker of the evolution of speech*, Wayne State University Press. \ Dan Dediu, Romanian, and Stephen Lavinson, Dutch, researchers at Max Planck Institute for Psychology, Germany, studied the mental processes involved in language production, comprehension and acquisition, and published their work *On the antiquity of language: the reinterpretation of Neanderthal linguistic capacities and its consequences*, Frontiers in Psychology, 4. \ Marius Albu, Romanian-born British multidisciplinary researcher, studied and expressed various forms of energy, e.g. cosmic, terrestrial, biological, human, cerebral including linguistic, and socioeconomic; and published the book entitled *Conventional and Non-conventional Forms of Energy*, United p.c., European Union. \ Sebastian Nordhoff, Harald Hammarström, Robert Forkel and Martin Haspelmath, German linguists, edited the work simply entitled *Tocharian*, Leipzig: Max Planck Institute for Evolutionary Anthropology.

5. Writing and printing

5.1. Pictography and proto-writing

Information or awareness gained through language, religion, music, exploration and inhabitation of new areas resulted in an unprecedented amount of knowledge that surpassed the capacity of oral transmission and reception of knowledge. At this stage, humans had to invent and develop drawings, engravings, paintings, pictographs, signs, methods etc. not necessarily to express their artistic skills, but mainly to make known their experience, feelings and thoughts by long lasting symbolic forms for remembering, transmitting and preserving their knowledge in the benefit of their relatives and descendants. Such symbolic forms of representations are exemplified in chronological order below.

40800 BP: *Paintings, engravings and symbols* in *Cueva de El Castillo* 'Cave of the Castle', Cantabria, Spain – apart from artistic representations (hand stencils and disks by blowing paint into the wall, and drawings of horses, stags, aurochs, goats, mammoths, dogs and other animals); there are distinctive *red, black* or *yellow dot clouds, rectangular forms, engraved* and *perforated baton, concentric circles,* other *geometric figures,* as well as *abstract and pictographic signs.*

40000-18000 BP: *Australian Aboriginal pictorial and pictographic representations* – including *dot painting* of animals, lakes and dreamtime images; megafauna such as *Genyornis; bark painting* and *aerial landscapes* on rock, sand or body painting; *rock carving* and *sculptural reliefs,* as well as *pictorial engraving.*

35600-18500 BP: *Painted* and *engraved mysterious signs* in *Altamira caves,* Cantabria, Spain – such as mono- and polychrome paintings of bisons, wild boars, horses, deer, anthropomorphic figures, handprints and hand stencils, created by using charcoal, ochre or haematite; and also three successive types of *drawings, signs,* and *pictographs.*

28000-10000 BC: *Painted* and *incised figures* and *non-figurative dot clusters* in *Lascaux caves,* Dordogne, France – cave paintings with mineral pigments or incised into the stone figures of equines, stags, bisons, felines, bird, bear, rhinoceros and human; and, more interestingly, *abstract signs, dot clusters, dots* and *lines of dots* as *pictographs.*

9500-8400 BC: *Decorations* at the archaeological site of *Göbekli Tepe,* Turkey – pillars with carved reliefs depicting   lions, bulls, boars, foxes, gazelles, donkeys, snakes   and other reptiles, arthropods, birds including vultures, as well as *humanoid figures* symbolizing venerated ancestors or anthropomorphic beings, and incised figures; and above all enigmatic *pictograms* representing *sacred symbols*.

8500-7000 BC: <u>*Rock paintings and symbols*</u> in *Bhimbetka rock shelters* and *caves*, Madhya Pradesh, India – stone age paintings, in green, dark red, then white and yellow, portraying bisons, tigers, rhinoceroses, buffaloes, antelopes, deer, peacock, and also human figures and hunting scenes, later added with *religious symbols*.

The above exemplified drawings, engravings, paintings, pictographs, signs and symbols were later developed into forms of *proto-writing*, as systems of *ideographic* and/or *mnemonic symbols* to convey information such as:

6600 BC: <u>*Jiahu symbols*</u> - consisting of *16 distinct markings* carved on tortoise shells, in Henan province, China, and belonging to the long process of sign-use development before the appearance of proper writing systems;

5300 BC: <u>*Tartaria tablets*</u>, three unbaked clay plates with incised *pictographic* or *writing-like symbols* only on one face, discovered in Tartaria village, 30 km from Alba Iulia, Romania, in the area of South-eastern Europe's *Vinča culture*, which are considered as maybe the earliest known form of writing in the world (N Vlasa, 1961).

Pictographic and proto-writing systems were in use for about two more millennia until the first scripts and proper writing were invented and applied.

5.2. Development of writing and printing

Script is defined as: handwriting as distinct from print; style of writing or a particular system of writing, e.g. cuneiform script; cursive writing and any system or style of writing; written characters or text; and also a manuscript or document.

Writing (English 'the act of one who writes or that which is written' from _write_ 'to express in writing'; Latin _scriptio_ 'act of writing, authorship, wording' from _scriptito_ 'to write often') emerged as a medium of communication representing language through the inscription of signs and symbols. In most general terms, writing is defined either as a method of recording information and is composed of _graphemes_, which may in turn be composed of _glyphs_; or systems of language representation through graphic means.

The result of writing is commonly called text, and the recipient of text is called reader. Motivations for writing include storytelling, correspondence, diary, and publication.

Much later, the _printing_ was invented and developed as a process for _reproducing text and images_, typically with ink on paper using a printing press, often carried out as a large-scale industrial process, representing an essential part of _publishing and transaction printing_.

Written communication evolved by three steps, considered to be 'information communication revolutions': _(i)_ use of pictographs by _pictograms_ made in stone, hence written communication was not yet mobile; _(ii)_ writing began to appear on paper, papyrus, clay, wax, etc., with common _alphabets_, and communication became mobile; and _(iii)_ transfer of information through controlled waves of _electromagnetic radiation_ (i.e. radio, microwave, infrared, and other _electronic signals_).

The process of production and dissemination of written texts, books, newspapers, and other means to inform the public is called _publishing_ that was later developed into Internet systems, micropublishing, websites, blogs, video game publishing, etc. The 1976 Copyright Act (the USA) defines _publication_ as follows:

"Publication is the distribution of copies or phonorecords of a work to the public by sale or other transfer of ownership, or by rental, lease, or lending. The offering to distribute copies or phonorecords to a group of persons for purposes of further distribution, public performance, or public display constitutes publication. A public performance or display of a work does not of itself constitute publication."

The notice for visually perceptible copies should contain three elements, namely - the symbol ©, or the word 'Copyright'; - the year of first publication of the work; and - the name of the owner of copyright in the work, or an abbreviation by which the name can be recognized, or a generally known alternative designation of the owner (e.g. © *2014 Thomas White*).

Writing and printing were developed by numerous inventions, discoveries, methods, systems, etc., from which a several dozens only are selected and displayed in chronological order below.

3400-3200 BC: *Ideographic or mnemonic symbols* first used as *pictographic characters* at Uruk, in southern Mesopotamia; *hieroglyphic characters* in Egypt, on the Nile Valley; and *symbolic characters* on the Indus Valley.

3200-3000 BC: *Sumerian cuneiform script* as the world's first *cuneiform writing* in southern Mesopotamia; *Proto-Elamite writing system* consisting of more than 1000 logographic signs, on clay tablets found in several sites of present-day Iran; *Egyptian hieroglyphic writing* on the Nile Valley (deciphered by JF Champollion, 1822-32); and *Indus script and pictographic writing* on the Indus Valley.

3000-2600 BC: *Coherent-written texts* in the first areas of civilization (Mesopotamia, Nile and Indus valleys).

2300-2000 BC: *Akkadian cuneiform writing* based on the Sumerian one, used by the Akkadian conquerors in the area of Tigris-Euphrates river system.

2000-1850 BC: *Semitic alphabetic script* invented by Semitic workers during their captivity in Egypt, and developed into *Proto-Sinaitic script* as the main source of all later alphabetic writings.

1750 BC: *Old Babylonian cuneiform script* derived from *Akkadian writing*, and then widely spread in the Middle East (D Charpin, 2010).

1750-1450 BC: *Cretan hieroglyphs* (syllabograms and logograms) used in Crete; and *Linear A script* earlier used in Aegean islands and Peloponnese, Greece (J-P Olivier, 1986; I Schoep, 1995; GA Owens, 1996; J-P Olivier *et al.*, 1996).

1550-1250 BC: *Hittite cuneiform writing* was a form of script adapted from a version of *Akkadian cuneiform* from northern Syria, which has been deciphered from the inscriptions discovered in ancient Hittite capital *Hattusa*, now Boğazköy in Turkey (B Hrozný, 1919).

1400-1050 BC: *Proto-Canaanite alphabet* descended from *Proto-Sinaitic script*, and evolved into the oldest consonantal *Phoenician* *alphabet* that was used in Syria and Palestine, and spread by Phoenician merchants across the Mediterranean world (SR Fischer, 2004; J-P Thiollet, 2005).

1375-1200 BC: *Linear B script* developed at Knossos in Crete, and at Pylos, Mycenae, Thebes and Tiryns in Greek mainland (AJ Evans, 1952; J Chadwick, 1987); about in the same time with *Hurrian* and later *Hittite writings* in Anatolia.

1250-1200 BC: *Hieroglyphic Hittite* was used in Anatolia; and *South Arabian alphabet* derived also from *Proto-Sinaitic script*, and later descended into *Arabic writing*.

c.**1200 BC**: *Early Chinese writing*, a body of writing in the *oracle bone script*, with logographic and phonetic representation, which was introduced during the late Shang Dynasty (Wu Ding, 1250-1192 BC).

1100-1046 BC: *Written History of China*, a *text from the Shang Dynasty* times using *Early Chinese writing*.

1000-500 BC: *Mesoamerican pictographic systems* developed first as *Olmec*, second as *Isthmian*, and third as *Zapotec*.

900-800 BC: *Aramaic alphabet* emerged from the former *Proto-Canaanite*, along with the *Hebrew alphabet*.

750 BC: *Greek alphabet* derived from the earlier *Phoenician alphabet*, and spread from Greek islands and mainland to southern Europe and North-eastern Mediterranean coasts (D Holton *et al.*, 1998).

700 BC: *Etruscan writing* was one of the *Tyrsenian written languages*, along with *Raetic* in the Alps and *Lemnian* in the Aegean island of Lemnos; which is known from c.13000 inscriptions, written from right to left, and developed as *Old Italic alphabet* that later was adopted by Romans (CE Pauli, 1885; OA Danielsson, 1907).

670-650 BC: *Latin (Roman) alphabet* evolved from a western variety of Greek alphabet, namely *Cumaean alphabet*, that was adopted and modified first by Etruscans who ruled the early Rome, and then by the ancient Romans to write their Latin language.

500-450 BC: *Arabic writing* was based on the previous *South Arabian alphabet* and spread through the Pre-Islamic Arabian world.

250 BC: *Mayan script*, a mixed logographic and syllabic writing system, was invented in Mesoamerica, as distinct of Olmec and Epi-Olmec *Isthmian script* (SD Houston *et al.*, 2000).

250-240 BC: *Brahmi script* was introduced and then widely used in the Indian Subcontinent.

150-100 BC: *Celtiberian script* developed from earlier *Celtic* and

144

Iberian identified on some inscriptions, and then used in North and central Iberian Peninsula.

AD 105: Invention of *paper and paper-making* in China (Ts'ai Lun, AD 97-105).

110-140: *Chinese dictionary* first written during Han Dynasty for understanding the Chinese writing system, on the basis of *seal script characters* and *xiaozhuan radicals* (Shuōwén Jiězi, 110-140).

150: *Runic alphabets* were first identified by the oldest *runic inscription* found in Denmark, and then used to write various German languages before the adoption of the Latin alphabet (R Blum, 1932; RI Page, 1999).

200-220: *Woodblock printing*, a technique for printing text, images or patterns on silk, in three colours, which originated during the Han Dynasty in China.

360-370: *Visigothic alphabet* was first used to translate the Bible from Greek into a *Germanic language* (Ulfilas, 350-380).

c.650: *Copper plate scrolls* used in Hindu-Malayan documents.

751: *Paper introduced to the Muslim world* after the Battle of Talas between the Arab Abbasid Caliphate and the Chinese Tang Dynasty, when the Arabs captured the paper and paper-making from the Chinese.

760-783: *German writing* known from fragments of an epic poem, called *Song of Hildebrand* dated c.760; and *Godesalc Evangelistary*, first using the script known as *Carolingian minuscule* (Godesalc, 781-783).

863-868: *Glagolitic and Cyrillic alphabets* introduced *Greek written signs* (from letters) to create *Slavic alphabets* representing the sounds of *Slavic languages*; being confirmed by the papal letter *Industriae tuae* from 880, the Cyrillic alphabet was used to translate The Bible and other texts into Slavic languages and then was spread by the Orthodox Church through large areas of Eastern and central Europe (St Cyril and Methodius, 863-880).

868: *Diamond Sutra* (Sanskrit *Vajracchedikā Prajñāpāramitā Sūtra*) is known by a copy translated into Chinese in 401, that was found among the *Dunhuang* manuscripts, and considered as the earliest complete survival of a date-printed book (now preserved in the British Library).

1040: *Movable type printing press* was first developed in China, as a system using clay and then porcelain (Bi Sheng, 1041-48).

1234: *Movable bronze type printing* was promoted during the Goryeo Dynasty in Korea, and then used for publishing 50 copies of the ritual book *Sangjeong Yemun* (Choe Yun-ui, 1230-41).

1298: _Wooden movable earthenware painting type_ was introduced during the Yuan Dynasty in China, and used for publishing the treatise _Nong Shu_ 'Book of Agriculture' (Wang Zhen, 1295-1313).

1377: _Oldest extant movable metal printed book_ was published in Korea and entitled _Jikji_ 'Selected Teachings of Buddhist Sages and Son Masters' (Bibliothèque Nationale de France, Paris).

1453: _Movable type printing technology_ originated in Germany by invention of a new _metal movable printing press_ and introduction of _oil-based ink_ that   is more durable than the previous water-based ink; which was an efficient printing process especially for European languages with their limited alphabets (JG Gutenberg, 1439-55).

1473: _First printed English book_ was a translation in English called _Recuyell of the Historyes of Troye_, achieved in Bruges, West Flanders (W Caxton, 1473-76).

1510-15: _Etching printing process_ was developed in Germany, using strong _acid_ or _mordant_ for cutting into unprotected parts on the surface of a _metal_ (copper, zinc, or steel) to create a design in _intaglio_ method of printmarking, which, along with engraving, represented the most important technique for _old master prints_ (D Hopfer, 1512-15; A Dürer, 1515-18).

1642: _Printmaking mezzotint process_ was developed as a _drypoint tonal method_, enabling half-tones to be produced on a plate roughened with thousands of little dots made by a metal tool with small teeth, where the ink was held when the face of the plate was wiped clean (L von Siegen, 1640-45).

1714: _Typewriter_ is a mechanical or electromechanical machine for writing in characters similar to those produced by printer's type, using keyboard-operated types striking a ribbon to transfer ink or carbon impressions onto the paper (H Mill, 1712-14).

1796: _Lithography printing technique_ resulted from the first _planographic process_ in printing by perfecting both the chemical process and the special form of printing press required for applying it on flat surface of a stone (JA Senefelder, 1792-1818).

1829: _Braille system of reading and writing_ was devised for use by blind or visually impaired people who can easily recognize _small cells_ as letters with a touch of finger (L Braille, 1824-37).

1837: _Chromatography_ was developed into a process providing consistently _high-quality results_ (G Engelmann, 1826-39).

1843: _Rotary printing press_ was provided with impressions curved around a _revolving cylinder_ to print on long continuous rolls of paper, thus working much faster than the old flatbed printing press (RM Hoe, 1843-47).

1938: *Xerography* or *electrophotography* was invented as a *dry photocopying technique* using no liquid chemicals (CF Carlson, 1938-59), later developed and marketed worldwide by the Xerox Corporation.

1969: *Laser printing* is an electrostatic *digital printing* process that rapidly produces high quality text and graphics by a laser beam passed over a charged drum, resulting in a differentially charged image and also using xerographic printing process (GK Starkweather, 1969-71).

1994-95: *Writing computer programs* such as sensory/modal preferences in *Collaborative writing systems*, and *ScriptSource* providing access to information on over 170 scripts used for the writing systems of thousands of languages (D Kaufer et al., 1995).

2000-02: *Electronic writing* represented by *(i) writing for asynchronous interpersonal communication* - in e-mail, mailing lists, newsgroups, and discussion groups; *(ii) writing for synchronous interpersonal communication* - in chatrooms, multi-user dimensions (MUDs), or multi-user dimensions object-oriented (MOOs); and *(iii) writing on the World Wide Web*, including hypertext (SP Ferris, 2002).

2004-06: *Digital printing* can be used at home, office, or engineering environment, either in *small format* for business offices and libraries, or *wide format* for drafting and design establishments, involving *printing technologies* such as blueprint, daisy wheel, dot-matrix, line printing, heat transfer, inkjet, electrophotography, laser, and solid ink printer; as well as *Digital contact printing* for *high-quality digital output* (Thomson Course Technology PTR Development, 2005).

2011: *Computer systems of writing* developed by *writing-programming relationship* (M Hartwig, 2011).

Literacy is the ability to *read* and *write*, comprising a complex set of capabilities to understand and use the dominant symbol system of a culture for personal and community development. *Visual literacy* includes in addition the capacity to understand visual forms of communication such as *body language, pictures, maps,* and *video*.

Besides the alphabetic and number systems, the use of media and electronic texts extends the concept of literacy in technological societies.

In general, the development of reading and writing implies rules of using *phonology* (awareness of speech sounds), *orthography* (spelling patterns), *semantics* (word meaning), *syntax* (grammar) and *morphology* (patterns of word formation).

The origin of literacy was related to numeracy and simplest arithmetic operations as early as around 8000 BC.

A definition of literacy was given by the UNESCO (United Nations Educational, Scientific and Cultural Organization) as "ability to identify, understand, interpret, create, communicate and compute, using printed and written materials associated with varying contexts. Literacy involves a continuum of learning in enabling individuals to achieve their goals, to develop their knowledge and potential, and to participate fully in their community and wider society".

Recent data indicate *literacy rates* ranging from 28 - 29% in Mali, Afghanistan, Burkina Faso and Niger; to 99 - 100% in Czech Republic, Denmark, France, Germany, Italy, Japan, UK, USA, Russia, South Korea, Canada and Norway; and in average of 84% for the entire world.

5.3. Timeline unfolding

The uneven development from pictographic and proto-writing to writing and printing systems can be displayed in an even manner, considering major stages delimited by the following transitional times

\ 39000 BP \ *Castillo pictographic signs*

\ 35000 BP \ *Australian engraving*

\ 31000 BP \ *Altamira drawings*

\ 27000 BP \ *Altamira signs*

\ 23000 BP \ *Altamira pictographs*

\ 18000 BC → 20015 BP \ *Lascaux signs and dots*

\ 15500 BC → 17515 BP \ *Lascaux pictographs*

\ 10000 BC → 12015 BP \ *Göbelki and Jiahu symbols*

\ 6000 BC → 8015 BP \ *Tartaria tablets, early writings*

\ 1900 BC → 3915 BP \ *Late writings and printings*

\ AD 1994 → 21 BP \ *Computer writing systems*

As the initial time of life was c.4200 million years BP, these transitional times can be inserted in the life evolution approximately such as:

\ 4200-0.039000 = 4199.961BP \

\ 4200-0.035000 = 4199.965BP \

\ 4200-0.031000 = 4199.969BP \

\ 4200-0.027000 = 4199.973BP \

\ 4200-0.023000 = 4199.977BP \

\ 4200-0.020015 = 4199.979985BP \

\ 4200-0.017515 = 4199.982485BP \

\ 4200-0.012015 = 4199.987985BP \

\ 4200-0.008015 = 4199.991985BP \

\ 4200-0.003915 = 4199.996085BP \

\ 4200-0.000021 = 4199.999979BP \ the present time.

However, the development of writing and printing can be properly integrated into the life evolution, on the basis of language timeline (see table by the end of subchapter 4.3) where the latest $N/10^4 = \delta$ sequence, delimited by values of argument \1.8805\...\1.8806\, includes $N/10^5 = \varepsilon$ sub-sequences suitable to unfold the course of writing and printing. As the final time would be $t_\bullet = 4400$ million years from the origin of life, the writing-printing sequences are delimited by transitional times $t_\varepsilon = t_\bullet \cdot tanh(\varepsilon)$, in million years from the same origin, as follows:

$$t_{1.88050} = t_\bullet \cdot tanh(1.88050) \approx (4400) \cdot (0.9545366) \approx 4199.96088;$$
$$t_{1.88051} = t_\bullet \cdot tanh(1.88051) \approx (4400) \cdot (0.9545375) \approx 4199.96479;$$
$$t_{1.88052} = t_\bullet \cdot tanh(1.88052) \approx (4400) \cdot (0.9545383) \approx 4199.96870;$$
$$t_{1.88053} = t_\bullet \cdot tanh(1.88053) \approx (4400) \cdot (0.9545392) \approx 4199.97261;$$
$$t_{1.88054} = t_\bullet \cdot tanh(1.88054) \approx (4400) \cdot (0.9545401) \approx 4199.97652;$$
$$t_{1.88055} = t_\bullet \cdot tanh(1.88055) \approx (4400) \cdot (0.9545410) \approx 4199.98043;$$
$$t_{1.88056} = t_\bullet \cdot tanh(1.88056) \approx (4400) \cdot (0.9545419) \approx 4199.98434;$$
$$t_{1.88057} = t_\bullet \cdot tanh(1.88057) \approx (4400) \cdot (0.9545428) \approx 4199.98825;$$
$$t_{1.88058} = t_\bullet \cdot tanh(1.88058) \approx (4400) \cdot (0.9545437) \approx 4199.99216;$$
$$t_{1.88059} = t_\bullet \cdot tanh(1.88059) \approx (4400) \cdot (0.9545446) \approx 4199.99607;$$
$$t_{1.88060} = t_\bullet \cdot tanh(1.88060) \approx (4400) \cdot (0.9545454) \approx 4199.99998.$$

Therefore, the timeline and sequences in development of writing and printing are finally presented as in the table below.

$N/10^5 =$ ε	Time (million years)		Sequences in development of writing and printing
	from origin $t_\varepsilon = t_\bullet \cdot tanh(\varepsilon)$	from present $t_\varepsilon - 4200$	
	4400	+200	
...	...	...	
1.88061	4200.00389	+0.00389 (AD5904)	–
1.88060	4199.99998	-0.00002 (AD1994)	_Computer writing systems
1.88059	4199.99607	-0.00393 (1916BC)	_ Late writings and printings
1.88058	4199.99216	-0.00784 (5826BC)	_Tartaria tablets, early writings
1.88057	4199.98825	-0.01175 (9736BC)	_Göbekli and Jiahu symbols
1.88056	4199.98434	-0.01566 (15660BP)	_ Lascaux pictographs
1.88055	4199.98043	-0.01957 (19570BP)	_Lascaux signs and dots
1.88054	4199.97652	-0.02348 (23480BP)	_Altamira pictographs
1.88053	4199.97261	-0.02739 (27390BP)	_Altamira signs
1.88052	4199.96870	-0.03130 (31300BP)	_Altamira drawings
1.88051	4199.96479	-0.03521 (35210BP)	_Australian engraving
1.88050	4199.96088	-0.03912 (39120BP)	_Castillo pictographic signs
...	...	...	
0	0	-4200	

References

1250-1192BC: Wu Ding, king of the Shang Dynasty in ancient China, became a historical figure after the *oracle script on bones* were unearthed at the ruins of his capital Yin, near modern Anyang, in AD1899.

AD97-105: Ts'ai Lun (Cai Lun), Chinese eunuch and political official, invented the *paper* (different to papyrus) and *paper-making process*; he was responsible for the first significant improvement and standardization of paper-making by adding essential new materials (bark, hemp, silk, fishing net) into its composition.

110-140: Shuōwén Jiězi, Han Dynasty Chinese linguist, analyzed the seal script characters, and introduced 540 *xiaozhuan* radicals, as the first *Chinese dictionary*, which was considered a major conceptual innovation in understanding the Chinese writing system.

350-380: Ulfilas (Wulfila), Cappadocian prelate, was consecrated a missionary bishop to the Visigoths, worked in Lower Moesia, and attended the Council of Constantinople in 360 in the interest of the Arian party; he first translated the Bible from Greek into a *Germanic language.*

781-783: Gadesalc, scribe commissioned by Charlemagne, King of the Franks, produced the magnificent book, now called *Gedesalc Evangelistary*, the first in which the script known as *Carolingian minuscule* appeared; the text being written with conventional capitals, but dedication with lower-case letters.

863-880: St Cyril and Methodius, (Greek Κύριλλος καί Μεθόδιος), the Apostle of Slaves, were Byzantine Greek Christian missionaries who worked among the Moravians; and the Tartar Khazars and Bulgarians of Thrace and Moesia, respectively; Methodius invented the *Glagolitic alphabet*, and then Cyril developed the *Cyrillic alphabet*, as the oldest Slavic alphabets created to translate The Bible and other texts from Greek into Slavic languages; the Cyrillic alphabet has been attributed to Cyril according to the papal letter *Industriae tuae* from 880.

1041-48: Bi Sheng, printer during Song Dynasty China, invented the first known *movable type printing press* technology; his system using Chinese porcelain.

1230-41: Choe Yun-ui, Korean civil minister during the Goryeo Dynasty, promoted the oldest extant *movable bronze type printing*, and together with other 16 scholars compiled the ritual book *Sangjeong Yemun*, that was printed with movable bronze type into 50 copies.

1295-1313: Wang Zhan, magistrate during the Yuan Dynasty of China, made innovations of the wooden movable printing system by an alternative method of baking *earthenware* printing type, and used it for publishing his treatise entitled *Nong Shu* 'Book of Agriculture'.

1439-55: Johannes Gensfleisch Gutenberg, German inventor of printing, began to print in Strasbourg achieving the construction of his *movable type printing press*, introducing of an *oil-based ink* more durable than the previous water-based inks, and using his new system to print *The Bible, Fragment of the Last Judgement*, and editions of Aelius Donatus's *Latin school grammar*.

1473-76: <u>William Caxton</u>, English merchant, diplomat, writer and printer, after joining the Flemish calligrapher Colard Mansion in Bruges, he first *printed in English* a book translated by himself and entitled *Recuyell of the Historyes of Troye*; returned in England, he set up a press at Westminster, that was used to print an edition of Geoffrey Chaucer's *The Canterbury Tales*.

1512-15: <u>Daniel Hopfer</u>, German artist, was the first to use *etching* in printmarking, his etching including an Augsburg's *horse armour* (German Historical Museum, Berlin), and *a sward* (Germanisches Nationalmuseum of Nuremberg).

1515-18: <u>Albrecht Dürer</u>, German painter and engraver, as an engraver on metal and a designer of woodcuts, he made metal plates with extreme finish and refinement, and woodcuts boldly drawn with a broad expressive line; his copperplates including the *Small Passion*, the *Knight, Death, and the Devil*, the *St Jerome in his Study*, and the *Melancholia*; whilst as the promoter of etching, he experimented on iron, and produced several plates in which all the lines are bitten with acid, such as *The Desperate Man, Man of Sorrows*, and *Agony in the Garden* (all in Metropolitan Museum of Art, New York).

1640-45: <u>Ludwig von Siegen</u>, German amateur artist, introduced a *tonal method* on a plate with thousands of little dots made by a metal tool with small teeth, where the ink was held the ink when the face of plate was wiped clean; his method developed as *mezzotint* was first used to print a portrait of *Countess Amalie Elisabeth of Hanau-Münzenberg*.

1712-14: <u>Henry Mill</u>, English inventor, had the idea to make a machine for putting letters on paper, and obtained patent for an apparatus 'for impressing or transcribing of letters singly or progressively one after another, so neat and exact as not to be distinguished from print, very useful in settlements and public records' as the first proposal for a *typewriter*.

1792-1818: <u>Johann Aloys Senefelder</u>, Bavarian actor, playwright and inventor, experimented with a novel *etching technique* using a greasy acid resistant ink to draw on a smooth fine-grained stone of *Solnhofen limestone*; discovered that this technique could be extended to allow printing on a flat surface of the stone alone as the first *planographic process* in printing; perfected both the chemical process and the special form of printing press required for using stones, resulting in *lithography*; established a publishing firm for lithography; and published his findings in the book *Vollstandiges Lehrbuch der Steindruckerei*.

1822-32: <u>Jean François Champollion</u>, French linguist and Egyptologist, using a copy of the *Rosetta stone*, and his acquired knowledge of the Coptic language, he was the first to decipher the Egyptian hieroglyphics, showing that the Egyptian writing system was a *combination of phonetic and ideographic signs*; and founding the *Egyptology*; his published works include *Panthéon égyptien, collection des personnages mythologiques de l'ancienne Égypte*; *Précis du système hiéroglyphique des anciens Égyptiens*; two-volume *Monuments de l'Égypte et de la Nubie*; *Grammaire égyptienne*; and

Dictionnaire égyptien en écriture hiéroglyphique; by which the studies of early Egyptian history and culture have been properly promoted.

1824-37: Louis Braille, French educationist and inventor, devised the *Braille system* of raised-point writing which blind or visually impaired people could both read and write by small cells capable of being recognized as letters with a single touch of a finger, and then published his main work *Method Of Writing Words, Music and Plain Songs*, which later was revised and republished.

1826-39: Godefroy Engelmann, French-German artist, brought the lithography to France, developed it into *chromolithography*, and was granted an English patent for his chromolithographic process that provided high-quality results; he produced numerous prints, including the plates for Baron Isidore Justin Séverin Taylor's celebrated collection of lithographs, called *Voyages pittoresques et romantiques dans l'ancienne France*.

1843-47: Richard March Hoe, US inventor, designed the *rotary printing press*, using a *revolving cylinder*, that worked much faster than the previous flatbed printing press; his invention was patented in 1847 and placed in commercial use in the same year.

1885: Carl Eugen Pauli, German scholar for Etruscan culture, published the reference work *Corpus Inscriptionum Etruscarum* 'Body of Etruscan Inscriptions'.

1907: Olof August Danielsson, Swedish linguist, continuing CE Pauli's work, he wrote *Corpus Inscriptionum Etruscarum / Vol.2*, Lipsiae: Barth.

1919: Bedřich Hrozný, Czech archaeologist and language scholar who deciphered *cuneiform Hittite*, wrote *Hethitische Keilschrifttexte aus Boghazköy* 'Hittite Cuneiform Inscriptions from Boğazkoy'*: In Umschrift, mit Übersetzung und Kommentar*, Leipzig: JC Hinrichs: Inscriptions, Hittite.

1932: Ralph Blum, US writer and cultural anthropologist, wrote *The Book of Runes - A Handbook for the use of Ancient Oracle: The Viking Runes*, Oracle Books, New York: St Martin's Press.

1938-59: Chester Floyd Carlson, US inventor, experimented with copying processes using *photoconductivity*, and discovered the basic principles of the electrostatic *xerography* (from Greek ξηρός 'dry' and γραφία 'writing'); after this process was patented, it was developed and marketed worldwide by the Xerox Corporation.

1952: Arthur John Evans, English archaeologist famous for unearthing the palace of Knossos in Crete, studied the *Linear B script* and published a work entitled *Scripta Minoa: the Written Documents of Minoan Crete; with Special Reference to the Archives of Knossos*, Vol.II.

1960-65: William Stokoe, US scholar, published *Sign Language Structure*, and co-authored *A Dictionary of American Sign Language on Linguistic Principles*, proving that American Sign Language fits the criteria for a natural language.

1961: Nicolae Vlasa, Romanian archaeologist, unearthed the *Tartaria tablets*, three unbaked clay plates covered with *pictographic writing* only on one face and dated 5300BC, which are considered the earliest known form of writing in the world; the tablets were found together with 26 clay and stone figurines and a shell bracelet, near the disarticulated bones of an adult male, in Tartaria village, 30 km from Alba Iulia, Romania.

1969-71: Gary Keith Starkweather, US engineer, working at Xerox in product development, he used a *laser* to draw an image directly onto the *copier drum*, and then adapted a Xerox 7000 copier to create Scanned Laser Output Terminal (SLOT).

1986: Jean-Pierre Olivier, French researcher of ancient Cretan scripts, wrote his article *Cretan Writing in the Second Millennium BC*, World Archaeology, 17(3).

1987: John Chadwick, British classical linguist, published his work *Linear B and Related Scripts*, Reading the Past, Third impression, University of California Press / British Museum.

1995: Ilse Schoep, Belgian researcher of *Cretan hieroglyphes*, published his work entitled *A New Cretan Hieroglyphic Inscription from Malia (MA/V Yb 03)*, Kadmos, 34. \ David Kaufer, Christine Neuwirth, Ravinder Chandhok and James Morris, US computer scientists, produced the work *Accommodating Mixed Sensory/Modal Preferences in Collaborative Writing Systems*, Computer Supported Cooperative Work (CSCW), 3, Kluver Academic Publishers, the Netherlands.

1996: Gareth Alun Owens, British-Greek academic, wrote the article *The Common Origin of Cretan Hieroglyphs and Linear A*, Kadmos, 35(2). \ Jean-Pierre Olivier, Louis Godard and Jean-Claude Poursat, French decipherers of *Linear A script*, published their work *Corpus Hieroglyphicarum Inscriptionum Cretae (CHIC)*, Études Crétoises, 31, De Boccard, Paris.

1998: David Holton, Peter Mackridge and Irini Philippaki-Warburton, British linguists specialized in ancient Greek inscriptions, made known the results of their work entitled *Garmmatiki tis ellenikis glossas*, Athens: Pataki.

1999: Raymond Ian Page, British historian, studied the English runes, and published his *An Introduction to English Runes*, Woodbridge: The Boydell Press.

2000: Stephen Douglas Houston, John Robertson and David Stuart, US anthropologists and Mayanists, made known their work *The Language of Classic Maya Inscriptions*, Current Anthropology, 41(3).

2002: Sharmila Pixy Ferris, US professor of interpersonal/communication studies, wrote her article *Electronic writing: Effects of computers on traditional writing*, The Journal of electronic publishing, 8(2), University of Michigan Press.

2004: Steven Roger Fischer, New Zealand historian of writing and reading, produced the books *A History of Writing*, and *A History of Reading (Globalities)*, both of them published by Reaktion Books Ltd., London.

2005: Jean-Pierre Thiollet, French writer and journalist, published his work *Je m'appelle Byblos* "My Name is Byblos', H & D Paris. \ Thomson Course

Technology PTR Development published *Mastering Digital Printing*, 2nd edition, a definitive guide to the world of *high-quality digital output*.

2010: <u>Dominique Charpin</u>, French historian of Mesopotamia, wrote a book in French that was translated in English as *Reading and Writing in Babylon*, Harvard University Press.

2011: <u>Michael Hartwig</u>, researcher at Multimedia University, Malaysia, published his work entitled *On the Relationship between Proof Writing and Programming: Some Conclusions for Teaching Future Software Developers*, in Software Engineering and Computer Systems, Springer Verlag.

6. Literature and its extent

6.1. Preliminary settings

In a large understanding, the term _literature_ (from Latin _literatura_ 'alphabet, grammar, writing formed with letters') usually refers to: written material such as poetry, novels, essays, etc., especially works of imagination characterized by excellence of style and expression and by themes of general or enduring interests \ writing in prose or verse regarded as having permanent worth through its intrinsic excellence \ imaginative or creative writing, especially of recognized artistic value \ the body of written works of a language, period, or culture \ written or printed matter of a particular type or on a particular subject \ printed material giving a particular type of information \ the body of written work produced by scholars or researchers in a given field of interest or activity \ language foregrounding literariness, as opposed to ordinary language \ art of composition in prose and verse \ literary and printed matter \ entire body of written works.

The major forms of literature are _novel_, _poem_, _drama_, _short story_, _novella_, and _myths_; which may be categorized into literary _genres_, determined by technique, tone or content, and represented by _epic_, _tragedy_, _comedy_, _tragi-comedy_, and _creative non-fiction_, in form of _prose_ or _poetry_, as well as _satire_, _allegory_, or _pastoral_. Other narrative forms include electronic literature; films, video and broadcast soap operas; as well as graphic novels and comic books.

Genres of literature are exemplified by _romance_, _satire_, _fable_, _mystery_, _crime_, _fantasy_, _erotica_, _adventure_, etc. Furthermore, the literature was differentiated into _fiction_ (fable, fairy tale, fantasy, folklore, historical fiction, horror, humour, legend, matafiction, mystery, mythology, poetry, realistic fiction, science fiction, short story, tall tale) and _non-fiction_ (biography or autobiography, essay, narrative non-fiction, speech, textbook, reference book).

Literature may also consist of text based on factual information, such as journalistic or non-fiction, including _polemical works_, _biographies_, _reflective essays_; or imaginative texts, such as fiction, poetry, or drama.

As a distinct matter, the prose literature referring to natural science, philosophy, history, law, and so on, as a distinct matter, is not included in the topic below.

Literary scholars and critics consider that the literary value of a text can be analyzed by phases such as: _(i)_ reading or re-reading the text with specific questions in mind; _(ii)_ additional reviewing of the text on communication basic ideas, events and names, depending on the complexity of text; *(iii)* thinking through the personal reaction to the text or book, by identification, enjoyment, significance, and application; *(iv)* identifying and considering most important ideas, the importance depending on context of class, assignment, and study guide; *(v)* returning to the text for location of specific evidence and passages related to the major ideas; *(vi)* using personal knowledge by principles of analyzing a passage - test, essay, research, presentation, discussion, or enjoyment.

Some literary scholars traced the re-evaluation process and outlined the evolving use of the very term 'literature', concluding that literature is 'a status rather than a quality'; and a text or a body of work is *literature* means that 'it is regarded, studied, read and analyzed in a literary way' (M Garber, 2011).

The study, evaluation and interpretation of literature are called *literary criticism*, which is comprehensively presented in the 2005 edition of *John Hopkins Guide to Literary Theory and Criticism*, John Hopkins University Press.

Literature is an effective means to communicate ideas, establishing bridges between writers as *senders* and readers as *receivers*. Communicative basis of literature have not always held a central place in the earlier literary studies, but became the core of many recent researches (R Sell, 2000).

Meanwhile, literature is a form of human expression, but not everything expressed in words can by counted as literature. The primarily informative writings, such as technical, scholarly, or journalistic, are excluded from the rank of literature by most critics, and only the forms of writing which possess artistic merit are considered as belonging to literature.

At different scales, a valuable literary work is like a 'temple' or 'cathedral' built up for present and future generations to be both accessible and impressive. The author of a proper literary work must have readily available components/elements and means to link them together into a unitary, harmonious and attractive creation for satisfying the spiritual needs and aspirations of his contemporaries and their descendants.

Both a writer and a critic of literary works would be better to keep in their mind some long lasting *Roman sayings*.

When doing a literary work, the *writer* may remember phrases or settings such as: *ex nihilo nihil fit* 'out of nothing, nothing comes' \ *facit indignatio versum* 'indignation makes verse' \ *poeta nascitur, non fit* 'the poet is born, not made' \ *ab imo pectore* 'from the bottom of the heart' \ *actum ne agas* 'do not do what is already done' \ *aut non tentaris aut perfice* 'either do not attempt or else achieve' \ *cave quid dicis, quando, et cui* 'beware what you say, when, and to whom' \ *causa sine qua non* 'an indispensable cause' \ *laborare est orare* 'work is prayer' \ *mirabile dictu* 'wonderful to tell' \ *da locum melioribus* 'give place to your betters' \ *age quod agis* 'do what you are doing' \ *callida junctura* 'a skilful connection' \ *simplex munditiis* 'elegant in simplicity' \ *crescit eundo* 'it grows as it goes' \ *esto perpetua* 'be lasting' \ *gradu diverso, via una* 'with different step on the one way' \ *festina lente* 'hasten gently' \ *non multa, sed multum* 'not many, but much' \ *ne quid nimis* 'nothing in excess' \ *a posse ad esse* 'from the possible to the actual' \ *limae labor* 'the labour of the file, or polishing' \ *littera scripta manet* 'what is written down is permanent' \ *nescit vox missa reverti* 'a word published cannot be recalled' \ *respice finem* 'look to the end' \ *finis coronat opus* 'the end crowns the work'.

When analyzing and evaluating a literary work, the <u>critic</u> may have in his mind sayings such as: *aut prodesse volunt aut delectare poetae* 'poets seek either to profit or to please' \ *arbiter elegantiae* 'judge of good taste' \ *auctor quae pretiosa facit* 'gifts that the giver adds value to' \ *experto crede* 'trust one who has tried' \ *facile est inventis addere* 'it is easy to add to things invented already' \ *errare est humanum* 'to err is human' \ *auxilium ab alto* 'help from on high' \ *verbum sapienti sat est* 'a word to the wise is enough' \ *circulus in probando* 'arguing in a circle, using the conclusion as one of the arguments' \ *quot homines, tot sententiae* 'as many men, so many opinions' \ *consensus facit legem* 'consent makes law or rule' \ *de gustibus non est disputandum* 'there is no disputing about tastes' \ *ex abusu non arguitur ad usum* 'from the abuse no argument is drawn against the use' \ *probatum est* 'it has been proved' \ *exceptio probat regulum* 'the exception proves the rule' \ *falsus in uno, falsus in omnibus* 'false in one thing, false in all' \ *habent sua fata libelli* 'books have their destinies'.

An attempt to count the literary works over a period of more than five millennia is a very difficult task, taking into account that the number of works written on papyrus scrolls kept in the Ancient Library of Alexandria (successively burnt between 48BC and AD391) was of around a half of million, and printed works continuously increased; for example in the USA only, the number of titles of all genres rose to about a quarter of million per year. Meanwhile, researches on strategies for literature reviewing show that the *qualitative searching strategies* are slow, hardly intensive and useful when the searched topics are not well conceptualized, involving the complexity of what has been previously written; whilst the *quantitative searching strategies* are relatively quick, cover a wide range of literature and a large amount of data, but lose the complexity.

World literature is usually regarded as either the sum total of the world's national literatures, or the circulation of works into a wider world beyond their countries of origin, and today is increasingly seen in a global context.

The development of world literature can be only pointed out by a limited number of written works subjectively chosen due to their historical, social, cultural, or educational importance; innovative genres, or styles; promotional movements, or forms of expression; distinctive quality, or elegance; etc.

6.2. Summary display of world literature

Literature emerged around five millennia ago and developed, on national tradition rather than on international course, by stages with various forms and genres, fiction and non-fiction, movements and authors, selectively displayed in chronological order below.

3050-1069 BC: *Pre-classical and classical Egyptian literature* − Funerary texts, epistles, poems, hymns, laments, narrative tales, stories, and histories, in both hieroglyphic and hieratic scripts; notably, *Narmer Palette* recording events from the time of unifying Egypt by the first pharaoh (Narmer, 3100-3000 BC); and *Book of Dead* (1550-1069 BC).

2150-2000 BC: *Sumerian-Akkadian epic poetry* − *Epic of Gilgamesh*, written on clay tablets, as the greatest surviving literary work, describing events related to the legendary 5th king of Uruk (Gilgamesh, 2600-2500 BC).

1700-1000 BC: *Pre-classical and classical Sanskrit literature* − Scholarly treatises, stories, and classical poetry, culminating with The *Vedas* epics, which consist of *Rig-veda* (1700-1100 BC), comprising several mythological and poetical accounts of the origin of world, hymns praising the gods, praying for life and prosperity, composed to be recited by the presiding priest; *Sama-veda* (1700-1000 BC), a collection of hymns and verses containing formulas to be sung by the priest who chants; *Yajur-veda* (1400-1000 BC), a series of hymns, learned knowledge, and formulas to be recited by the officiating priest; and *Atharva-veda* (1100-1000 BC), a collection of spells, incantations, apotropaic charms and speculative hymns, composed for correct application of the procedures of rituals and heals.

770-200 BC: *Classical Chinese texts* − Woodblock printing from the Eastern Zhou Dynasty, recording religious and philosophical texts, military science, history, books of poems and songs, such as collection of great philosophical writings contained in the *Hundred Schools of Thought* (770-220 BC).

760-230 BC: *Classical Greek literature* − *Epic poetry* (Homer, 760-710 BC), *lyric poetry* (Pindar, 498-440 BC), *tragedy* (Sophocles, 468-401 BC), *tragic drama* (Euripides, 431-406 BC), *comic drama* (Aristophanes, 424-388 BC), *sceptic literature* (Timon of Phlius, 285-230 BC).

600 BC-AD 1380: _Classical Persian literature_ ‾ _Religious_ and _philosophical prose_ (Zarathustra, 600-553 BC, and his followers), _poetry_ (Firdausi, AD 980-1010; Hafez-e Shirazi, 1360-80).

330-100 BC: _Hellenism_ ‾ Movement promoted and spread by Greek speaking authors during and after the Macedonian expansion, e.g. _Egyptian literature written in Greek language_ (Manetho, 250-230 BC).

166 BC-AD 180: _Classical Roman literature and oratory_ ‾ _Drama_ (P Terentius, 166-160 BC), _climax of oratory_ (MT Cicero, 80-43 BC), _prose_ (MT Varro, 70-27 BC), _culmination of literature_ (Virgil, 37-19 BC; Q Horace, 35-15 BC; S Propertius, 22-15 BC; Ovid, 15 BC-AD 17), _imperial literature_ (LA Seneca, AD 45-65; MF Quintilianus, 70-95; Silius Italicus, 80-100), _literature_ and _rhetoric_ (L Apuleius, 160-180).

AD 155-220: _Greek satire, rhetoric and anecdotage_ ‾ _Religious_ and _historical writings_ (Lucian, 155-170), _rhetoric_ and _anecdotes_ (Athenaeus of Naucratis, 180-220).

355-410: _Late Latin and Greek literature_ ‾ _Epistles_ and _satires_ (FC Julian, 355-363), _religious treatises_ (Gregory of Nyssa, 380-394), _poetry_ (MAC Prudentius, 380-410), _religious literature due to Syrians_ (J Chrysostom, 385-407).

540-1148: _Byzantine literature_ – _Historical writings_ (Procopius of Caesarea, 540-565), _religious works_ (John of Damascus, 720-745), _text-books_ (Anna Comnena, 1137-48).

580-600: _Pre-Islamic Arabic literature_ – _Prose_ and _poetry_ (Antar, 580-600).

690-736: _Classical Japanese literature_ – _Great anthology of poetry_ (Kakinomoto no Hitomaro, 690-710; Yamabe no Akahito, 724-736), _collections of poetic pieces_ including _Kojiki_ (712), _chronicle_ such as _Nihon Shoki_ (720).

948-1864: _Icelandic literature_ ‾ _Poetry_ (E Skallagrímsson, 948-980), _tales_ and _legends_ (J Árnason, 1862-64).

1090-1773: _Ukrainian and Russian literature_ ‾ _Chronicle_ (Ukrainian Nestor the Chronicler, 1090-1114), _epic_ and _narrative literary works_ (Russian Sophonia of Ryazan, 1400-20; A Nikitin, 1469-72), _poetry_ and _plays_ (VK Trediakovsky, 1735-52; IS Barkov, 1762-73).

1170-1470: _Early French literature_ – _Romances_ (Chrétien de Troyes, 1170-90), _poetry_ (F Villon, 1463-70).

1175-1560: _Early Serbian and Croatian literature_ – _Gospel books_ (Serbian Gligorije the Pupil, 1175-86), _Latin religious writings_

(Croatian M Marulić, 1490-1520), *novels* (Croatian P Zoranić, 1536-60).

1195-1218: *Early German literature* – Stories and *poems* (Austrian W von der Vogelweide, 1195-1230), *lyrics* and *epic stories* (German W von Eschenbach, 1200-18).

1250-1659: *Flemish-Dutch literature* ⁻ *Prose* (Beatrijs van Nazareth, 1250-65), *poetry* (J van Maerlant, 1263-84), *dramatic works* (J van den Vondel, 1654-59).

1307-1471: *Italian literary language and early literature* – Prose in Latin and Italian (Dante Alighieri, 1307-20), *poetry* and *prose* (F Petrarch, 1341-70), *mythological works* and *treatises* (G Boccaccio, 1350-70), *humanist works* (L Valla, 1440-71).

1350-1650: *Renaissance* – Revival of letters and art as a transition from the *mediaeval* to the *modern world*, in Europe, under the influence of Greek and Latin literatures (e.g. F Petrarch, 1341-70; G Boccaccio, 1350-70; L Ariosto, 1506-32; L de Camoëns, 1558-72; T Tasso, 1562-93; M de Cervantes, 1585-1605; C Marlowe, 1588-93; W Shakespeare, 1590-1613). \ *Humanism* ⁻ Literary culture and classical studies that puts human interests and mind paramount, rejecting supra-natural or belief, and applying *pragmatism, critical thinking* and *evidence* (rationalism, empiricism) over established doctrine or faith (fideism), in all sciences and arts, including linguistics (e.g. L Valla, 1440-71; F Rabelais, 1532-53; ME de Montaigne, 1569-88; B Jonson, 1606-14; J Milton, 1643-71).

1369-1460: *English literary language and early literature* – Tales and *legends* (G Chaucer, 1369-87), *chronicle* (J Capgrave, 1430-60).

1395-1774: *Early Scottish literature* – Chronicle (Andrew of Wyntoun, 1395-1410), *poems* in Scots and English (R Fergusson, 1771-74).

1450-1787: *Italian literature* – Florence's first library and *lyric poetry* (C de' Medici and L de' Medici, 1450-90), *Renaissance style-comedies* and *poems* (L Ariosto, 1506-32), *poems* and *orations* (MG Vida, 1527-37), *verse* and *prose* (T Tasso, 1562-93), *plays* (C Goldoni, 1753-87).

1512-1743: *Early Romanian literature* – Philosophical and *ethic writings* (Wallachian N Basarab, 1512-21), *earlier chronicles* (Moldavian G Ureche, 1620-47; M Costin, 1672-87), *poetry* (Moldavian D Barila, 1673-90), *historical works* (Moldavian N Costin, 1675-1711), *popular* and *religious writings* (Wallachian A Ivireanul, 1700-08), *political* and *historical writings* (Wallachian C Cantacuzino, 1700-16), *later chronicles* (Moldavian I Neculce, 1719-43).

1515-72: *Early Portuguese literature and literary language* – Plays and *farces* (G Vicente, 1515-35), *Renaissance style-poetry* (L de Camoëns, 1558-72).

1532-1817: *French literature* – *Renaissance style-works* (F Rabelais, 1532-53), *sceptical-philosophical writings* and *essays* (ME de Montaigne, 1569-88), *comedies* and *tragedies* (P Corneille, 1638-74), *romances* (M de Scudéry, 1641-67), *ballads* and *fables* (J de La Fontaine, 1654-84), *comic works* and *masterpieces* (JBP Molière, 1659-73), *poetry* and *plays* (J Racine, 1664-91), *great dramas* and *poems* (FMA Voltaire, 1718-78), *tragedies* (PJ de Crébillon, 1742-55), *plays* and *comedies* (PAC de Beaumarchais, 1767-84), *novels* and *romances* (M de Staël, 1789-1817).

1579-1781: *English literature* – *Poetry* and *prose* (E Spenser, 1579-96), *poems* (T Lodge, 1580-95), *lyric works* (T Watson, 1582-93), *verse* (W Warner, 1585-1606), *tragedies* and *plays* (C Marlowe, 1588-93), *great Renaissance plays* and *comedies* (W Shakespeare, 1590-1613), *tragedies* and *poems* (G Chapman, 1595-1618), *tragedies* and *plays* (J Marston, 1598-1607), *poems* and *prose* (J Donne, 1598-1617), *comedies* and *tragedies* (T Dekker, 1600-24), *satires* and *comedies* (B Jonson, 1606-14), *poetry* and *tragedies* (T Heywood, 1607-37), *plays* (J Fletcher, 1611-24), *tragedies* (J Webster, 1612-23), *poems* and *prose* (F Quarles, 1620-44), *pastorals* and *poems* (G Wither, 1622-41), *prose* and *poems* (J Milton, 1643-71), *prose* and *biographies* (I Walton, 1653-78), *verse plays* and *comedies for stage* (J Dryden, 1658-99), *comedies* and *plays* (W Wycherley, 1671-77), *prose* (J Vanbrugh, 1696-1726), *satires* and *verse* (J Swift, 1704-58), *satires* and *major fictions* (D Defoe, 1719-28), *tragedies* and *poems* (E Young, 1719-45), *novels* (S Richardson, 1740-54), *apologies* and *adventure writings* (H Fielding, 1741-51), *poems* and *fables* (S Johnson, 1749-81).

1585-1680: *Spanish literary language and literature* – *Renaissance romances* and *poems* (M de Cervantes, 1585-1605), *poetry* and *epic writings* (L de Vega, 1602-34), *poems* in affected style (L de Góngora, 1613-25), *dramas* and *sacramentals* (PC de la Barca, 1635-80).

1650-1755: *Afghan and Kurdish poetry* – *Afghan-Pashtun* (KK Khattak, 1650-80), *Kurdish* (K Qubadi, 1735-55).

1700-1800: *Enlightenment* – Cultural movement emphasizing *reason* and *individualism* rather than tradition, intending to reform society using reason, to challenge ideas based on tradition and faith, and to advance knowledge through the *scientific method*, promoting scientific thought, scepticism, and intellectual interchange (e.g. FMA Voltaire, 1718-78; JW von Goethe, 1767-1832; JG Herder, 1778-1800; JCF

Schiller, 1781-1804).

1750-1850: _Romanticism_ – Artistic and literary movement that was partly a reaction to the Industrial Revolution, a revolt against the aristocracy, and also a reaction against the scientific _rationalization of nature_ (e.g. R Burns, 1786-96; GGN Byron, 1809-24; J Austen, 1811-17; J Keats, 1816-20).

1759-1827: _Irish literature_ – Poetry and _plays_ (O Goldsmith, 1759-73), _poetry_ and _novels_ (T Moore, 1807-27).

1767-1912: _German literature_ – Drama masterpieces and _novels_ (JW von Goethe, 1767-1832), _folksongs_ and _poetry_ (JG Herder, 1778-1800), _poems_ and _great historical dramas_ (JCF Schiller, 1781-1804), _novels_ and _elegies_ (F Hölderlin, 1795-1802), _tales_ and _plays_ (H von Kleist, 1800-11), _tales_ and _sagas_ (JL Carl Grimm and W Carl Grimm, 1812-37), _poems_ and _prose_ (H Heine, 1812-54), _tales_ (ETW Hoffmann, 1814-22), _plays_ and _novels_ (GJR Hauptmann, 1889-1912).

1768-90: _Swedish literature_ – Collections of poems and _epistles_ (CM Bellman).

1774-1851: _Serbian literature_ – Prose and _short stories_ (D Obradović (1774-1810), _poetry_ (P Petrović-Njegoš, 1833-51).

1786-1827: _Scottish literature_ – Poems and _songs_ (R Burns, 1786-96), _novels_ and _tales_ (W Scott, 1802-27).

1792-1923: _Russian literature_ – Poetry (NM Karamzin, 1792-1826), _poems_ and _plays_ (FN Glinka, 1815-30), _outstanding poems_ and _novels_ (AS Pushkin, 1820-30), _novels_ and _dramas_ (NV Gogol, 1831-42), _novels_ and _poetry_ (MY Lermontov, 1837-41), _plays_ and _novels_ (IS Turgenev, 1850-70), _trilogy_ and _great novels_ (LN Tolstoy, 1851-99), _poems_ and _epic writings_ (NA Nekrasov, 1858-78), _great philosophical novels_ (FM Dostoevsky, 1866-80), _dramas_ and _short stories_ (AP Chekhov, 1886-1916), _novels_ and _plays_ (M Gorky, 1895-1923).

1795-1904: _English literature_ – Poems and _biographies_ (R Southey, 1795-1847), _lyrical ballads_ and _poems_ (W Wordsworth, 1798-1840), _satires_ and _poems_ (GGN Byron, 1809-24), _novels_ (J Austen, 1811-17), _lyrical poetry_ and _prose_ (PB Shelley, 1813-22), _sonnets_ and _romances_ (J Keats, 1816-20), _great novels_ (CJH Dickens, 1837-64), _novels_ (WM Thackeray, 1840-60; (EJ Brontë, 1845-47), _novels_ and _stories_ (C Brontë, 1847-54), _poems_ and _verse novelettes_ (A Tennyson, 1850-89), _dramas_ and _novels_ (AC Swinburne, 1865-94), _poetry_ (W Morris, 1872-96), _poems_ and _dramas_ (T Hardy, 1878-1904).

1801-55: _Danish literature_ – Poetry and _plays_ (AG Oehlenschläger,

1801-47), *novels* and *fairy tales* (HC Andersen, 1835-55).

1801-1914: *French literature* – *Prose* and *memoirs* (R de Chateaubriand, 1801-48), *poems* and *histories* (AML de Lamartine, 1820-51), *dramas* and *great novels* (VM Hugo, 1822-82), *great novels* (H de Balzac, 1829-33; HMB Stendhal, 1830-39), *poems* and *novels* (T Gautier, 1830-69), *novels* (George Sand, 1832-61), *great novels* and *plays* (A Dumas, 1836-50), *novels* and *comedies* (E Labiche, 1838-60), *poems* (CP Baudelaire, 1857-69), *novels* (G Flaubert, 1857-80; P Mérimée, 1860-81; E de Goncourt and J de Goncourt, 1860-87), *theatrical pieces* and *novels* (A Daudet, 1862-90), *well-known novels* (J Verne, 1863-73), *stanzas* and *poems* (Sully-Prudhomme, 1865-88), *great novels* (É Zola, 1867-1903), *romances* (P Verlaine, 1869-93), *symbolist poems* (S Mallarmé, 1875-99), *novels* and *satirical works* (A France, 1879-1914), *novels* (G de Maupassant, 1880-92).

1807-1923: *Austrian literature* – *Dramatic poetry* (FS Grillparzer, 1807-38), *lyric poetry* (RM Rilke, 1900-23).

1808-1916: *US literature* – *Satirical essays* and *tales* (W Irving, 1808-37), *stories* of sea and of Native American Indians (JF Cooper, 1821-45), *verse* and *detective stories* (EA Poe, 1827-45), *novels* and *stories* (N Hawthorne, 1828-63), *anti-slavery novels* (HB Stowe, 1834-70), *novels* and *stories* (H Melville, 1846-57), *poems* (W Whitman, 1855-91), *sketches* and *novels* (M Twain, 1872-84), *novels* and *stories* (H James, 1875-1916).

1813-2008: *Vietnamese literature* – *Poetry* (Nguyen Du, 1813- 20), *poems* and *novels* (Linh Dinh, 2000-08).

1818-1910: *Polish literature* – *Poems* and *dramas* (AB Mickiewicz, 1818-50), *novels* and *great trilogy* (H Sienkiewicz, 1884-1910).

1820-2005: *Portuguese literature* – *Romantic poetry* and *poems* (A Garrett, 1820-53), *realist style novels* (JM de Eça de Queiroz, 1870-1900), *novels* and *allegories* (J de Sousa Saramago, 1982-2005).

1821-1990: *Thai literature* – *Poetry* (Sunthom Phu, 1821-55), *novels* and *romantic works* (Kulap Saipradit, 1928-37), *novels* and *short stories* (Kukrit Pramoj, 1951-90).

1822-1965: *Greek literature* – *Hymns* and *poems* (D Solomos, 1822-47), *poetry* (G Seferis, 1931-65).

1829-1936: *Norwegian literature* – *Poems* and *plays* (HAT Wergeland, 1829-44), *dramas* (H Ibsen, 1850-99), *novels* (K Hamsun, 1890-1936).

1830-57: *Filipino literature* – *Poems* and *comedies* (F Balagtas).

1832-1977: _Czech literature_ ‒ _Romantic poetry_ and _sketches_ (KH Mácha, 1832-36), _poems_ and _patriotic writings_ (J Seifert, 1921-77).

1840-45: _Ukrainian literature_ ‒ _Poetry_ and _folk-tales_ (TH Shevchenko).

1842-1948: _Hungarian literature_ – _poems_ and _novels_ (S Petöfi).

1847-56: _Brazilian literature_ ‒ _Poems_ and _plays_ (AG Dias).

1848-70: _Finnish literature_ ‒ _Tales_ and _poems_ (JL Runeberg).

1849-1991: _Spanish literature_ – _Romances_ (F Caballero, 1849- 56), _poems_ (FG Lorca, 1932-47), _novels_ and _short stories_ (CJ Cela, 1942-91).

1850-Today: _Existentialism_ ‒ Movement sharing the belief that philosophical thinking begins with the human subject, characterized by the _existential attitude_ (e.g. FM Dostoevsky, 1866-80; F Kafka, 1913-24; J-P Sartre, 1938-57; A Camus, 1942-53; É Ionesco, 1950-80).

1865-1952: _Romanian literary language and literature_ – _Historical works_ and _dramas_ (BP Hasdeu, 1865-92), _great lyric verse_ and _prose_ (M Eminescu, 1870-83), _plays_ and _dramas_ (IL Caragiale, 1879-1910), _novels_ and _poems_ (D Zamfirescu, 1883-1919), _historical trilogy_ and _prose_ (BS Delavrancea, 1886-1912), _ballads_ and _pastorals_ (G Cosbuc, 1893-98), _symbolist poems_ (G Bacovia, 1898-1934), _novels_ (M Sadoveanu, 1902-52), _ballads_ and _poems_ (G Toparceanu, 1905-37), _masterpiece poems_ and _novels_ (T Arghezi, 1907-50), _poems_ and _prose_ (I Minulescu, 1909-31), _novels_ and _plays_ (L Rebreanu, 1912-38).

1867-76: _Bulgarian literature_ ‒ _Poems_ and _patriotic writings_ (H Botev).

1872-2004: _Swedish literature_ ‒ _Dramas_ and _novels_ (A Strindberg, 1872-1907), _novels_ (SOL Lagerlöf, 1891-1918), _condensed poetical writings_ (TG Tranströmer, 1972-2004).

1877-2013: _Japanese literature_ ‒ _Novels_ and _legends_ (Koizumi Yakumo, 1877-1904), _stories_ and _novels_ (Yasunari Kawabata, 1922-54), _novels_ (Oë Kenzaburo, 1958- 85), _novels_ and _plays_ (Abe Kobo, 1959-80), _short stories_ and _novels_ (Kenzaburō Ōe, 1964-2013).

1878-1948: _Indian literature_ – _Poetry_ and _philosophic prose_ (R Tagore, 1878-1940), _autobiographical writings_ and _collected works_ (M Gandhi, 1927-48).

1879-2009: _Italian literature_ ‒ _Novels_ and _tragedies_ (G d'Annunzio, 1879-1904), _dramas_ and _novels_ (L Pirandello, 1904-30), _plays_ and _poetry_ (E De Filippo, 1931-64), _poems_ (S Quasimodo, 1942-65), _plays_ and *comedies* (Dario Fo, 1958-2009).

1879-2010: *Irish literature* – *Novels* and *vampire story* (B Stoker, 1879-1911), *poems* and *tales* (WB Yeats, 1885-1936), *classic children's stories* and *novels* (OF Wilde, 1888-94), *novels* (E Somerville, 1889-1925), *dramas* and *plays* (GB Shaw, 1889-1934), *novels* (JAA Joyce, 1907-22), *novels* and *plays* (SB Beckett, 1938-70; Dame I Murdoch, 1957-93), *poems* and *plays* (SJ Heaney, 1966-2010).

1883-1916: *Belgian literature* – *Poems* and *plays* (E Verhaeren, 1883-1916), *dramas* and *poetry* (M Maeterlinck, 1889-1910).

1883-2012: *Canadian literature* – *Poetry* and *collection of poems* (EP Johnson, 1883-1912), *short stories* (AA Munro, 1968-2012).

1886-1962: *English literature* – *Verse* and *stories* (R Kipling, 1886-1936), *novels* and *stories* (HG Wells, 1895- 1945), *stories* and *novels* (J Galsworthy, 1897-1929), *novels* and *essays* (V Woolf, 1912-41; AL Huxley, 1921-62).

1887-1926: *Australian literature* – *Poems* and *prose* (H Lawson, 1887-1908), *poems* and *fiction writings* (IMD Mackellar, 1904-26).

1887-1988: *Scottish literature* – *Historical romances* and *modern detective stories* (AC Doyle, 1887-1906), *stories* and *novels* (Dame MS Spark, 1951-88).

1895-1946: *French literature* – *Poems* and *aesthetic studies* (P Valéry, 1895-1951), *novels* (APG Gide, 1902-50), *poems* and *novels* (J Romains, 1908-46), *novels* (RM du Gard, 1909-41), *poetry* and *plays* (G Apollinaire, 1909-46), *novels* (R Rolland, 1910-33), *prose* (M Proust, 1913-22), *prose* and *poems* (M Jacob, 1917-44).

1899-2004: *Hungarian prose* – *Symbolist-style poems* (E Ady, 1899-1912), *novels* and *factual writings* (I Kertész, 1988-2004).

1899-2005: *Polish literature* – *Novels* and *historical writings* (WS Reymont, 1899-1919), *poetry* and *essays* (W Szymborska-Włodek, 1981-2005).

1903-38: *Pakistani literature* – *Poetry* and *philosophical writings* (AM Iqbal).

1903-92: *US literature* – *Novels* (J London, 1903-13), *poems* and *novels* (WC Williams, 1909-62), *lyric poetry* (RL Frost, 1913-62), *poetry* and *dramas* (TS Eliot, 1920-58), *novels* and *stories* (EM Hemingway, 1923-52), *plays* (EG O'Neill, 1926-59), *novels* (W Faulkner, 1929-59), *poetry* (RG Eberhart, 1930-76), *novels* (JE Steinbeck, 1935-61; T Morrison, 1970-92).

1904-55: _Swiss literature_ – Novels and _poetry_ (H Hesse).

1907-43: _Welsh literature_ – Poetry and _collection of poems_ (WH Davies).

1909-20: _Futurism_ – Movement glorifying the war, machine age-speed, and also a sort of 'dynamism' as a revolt _against tradition_ (EFT Marinetti).

1910-97: _Mexican literature_ – Poetry (RL Velarde, 1910-21), _poetry_ and _essays_ (OP Lozano, 1956-97).

1910-65: _Modernism_ – Movement characterized by a _self- conscious break_ with traditional styles of poetry and verse (e.g. P Valéry, 1895-1951; V Woolf, 1912-41; BF Brecht, 1918-45; T Mann, 1918-55; EG O'Neill, 1926-59; W Faulkner, 1929-59).

1910-90: _Russian-Soviet literature_ – Novels and _short stories_ (IA Bunin, 1910-46), _poems_ and _satirical plays_ (VV Mayakovsky, 1918-30), _prose_ and _series of novels_ (IG Ehrenburg, 1921-66), _lyric poetry_ and _political poems_ (BL Pasternak, 1927-57), _masterpiece novels_ (MA Sholokhov, 1928-69), _novels_ and _factual accounts_ (AI Solzhenitsyn, 1956-90).

1911-2004: _Turkish literature_ – Poems and _national anthem_ (MA Ersoy, 1911-36), _novels_ and _screenplays_ (FO Pamuk, 1990-2004).

1913-2003: _Austrian literature_ – Novels (F Kafka, 1913-24), _biographies_ and _psychological novels_ (S Zweig, 1924-42), _novels_ and _plays_ (E Jelinek, 1975-2003).

1916-20: _Dadaism_ – Literary-artistic movement that was opposite to war and cultural values, but espoused _anarchic individualism_ and freedom from artistic convention (H Ball, H Arp and T Tzara).

1918-70: _Yugoslav literature_ – Novels and _short stories_ (I Andrić).

1918-95: _German literature_ – Plays and _poems_ (BF Brecht, 1918-45), _novels_ (T Mann, 1918-55), _novels_ and _dramas_ (L Feuchtwanger, 1923-54), _novels_ (H Böll, 1949-75; GW Grass, 1959-95).

1918-2002: _French literature_ – Novels and _biographies_ (A Maurois, 1918-59), _novels_ and _plays_ (F Mauriac, 1922-38), _dramatic novels_ (A Malraux, 1928-51), _novels_ and _detective stories_ (GJC Simenon, 1931-60), _dramas_ and _novels_ (J-P Sartre, 1938-57), _novels_ and _plays_ (A Camus, 1942-53), _novels_ and _poems_ (J Genet, 1944-49), _novels_ and _plays_ (F Sagan, 1954-83), _novels_ (C Simon, 1958-2002).

1918-2006: _Peruvian literature_ – Poetry and _plays_ (CAV Mendoza,

1918-38), *prose* and *essays* (JMPV Llosa, 1962-2006).

1918-2012: <u>*Chinese literature*</u> ⁻ *Short stories* and *story cycles* (Lu Xun, 1818-25), *major trilogy* (Ba Jin, 1931-40), *plays* and *novels* (Gao Xingjian, 1983-2012), *novels* and *short stories* (Mo Yan, 1987-2005).

1920-57: <u>*Venezuelan literature*</u> ⁻ *Novels* and *patriotic writings* (RÁ del MC Gallegos).

1923-80: <u>*French literature due to Romanians*</u> – *Prose* (Panait Istrati, 1923-30), *drama-plays* 'Theatre of the Absurd' and *novels* (E Ionesco, 1950- 80).

1924-66: <u>*Chilean literature*</u> ⁻ *Love poems* and *odes* (P Neruda).

1925-34: <u>*Korean literature*</u> ⁻ *Series of poems* (Kim Sowol).

1927-83: <u>*Egyptian literature*</u> ⁻ *Tragedies* and *poetry* (A Shawqi, 1927-31), *novels* (N Mahfouz, 1947-83).

1930-91: <u>*Romanian literature*</u> – *Novels* and *philosophical writings* (C Petrescu, 1930-40), *novels* and *plays* (M Sebastian, 1933-47), *novels* (M Preda, 1955-67), *poetry* (N Stanescu, 1972-82), *religious* and *philosophical writings* (IP Culianu, 1975-91).

1931-68: <u>*Icelandic literature*</u> ⁻ *Novels* and *epic writings* (HK Laxness).

1931-2007: <u>*English literature*</u> – *Verse* and *poems* (J Betjeman, 1931-76), *tragedies* and *romances* (CA Cookson, 1955-90), *poetry* (T Hughes, 1956-94), *plays* and *screenplays* (H Pinter, 1957-2007), *detective stories* (PD James, 1962-97), *novels* and *television pieces* (L La Plante, 1972-2005).

1932-41: <u>*Indonesian literature*</u> ⁻ *Lyrical prose* and *collections of poems* (TA Hamzah).

1935-92: <u>*Swiss and British literature due to Bulgarians*</u> ⁻ *Novels* and *memoirs* (E Canetti).

1949-2003: <u>*South African literature*</u> ⁻ *Prose* (N Gordimer, 1949-95), *novels* and *essays* (JM Coetzee, 1977-2003).

1952-61: <u>*French West Indian literature*</u> ⁻ *Revolutionary philosophical writings* (FO Fanon).

1953-80: <u>*Sudanese literature*</u> ⁻ *Patriotic poetry* (GA Rahman).

1953-2002: <u>*Lithuanian-Polish literature*</u> ⁻ *Prose* and *poetry* (C Miłosz).

1954-2003: <u>*British literature*</u> – *Stories* and *plays* (R Dahl, 1954-90), *dramas* and *radio plays* (T Stoppard, 1964-98), *novels* and *plays* (DM

Lessing, 1950-2003).

1958-2002: *Nigerian literature* ⁻ *Novels* and *poetry* (C Achebe, 1958-87); *plays* and *poetry* (AO Soyinka, 1976-2002).

1961-87: *British literature due to Trinidadians* ⁻ *Novels* and *essays* (VS Naipaul).

1963-2004: *French-Mauritian literature* ⁻ *Novels* and *short stories* (J-M G Le Clézio).

1965-80: *Ethiopian literature* ⁻ *Plays* and *dramatic poems* (T Gabre-Medhin).

1965-2008: *Moldovan literature* ⁻ *Representative poems* and *songs* (G Vieru).

1965-Today: *Post-Modernism* ⁻ Movement characterized by heavy reliance on *techniques* like fragmentation, paradox, and questionable narrators, and defined as a *style* or *trend* emerged in the post Second World War era, as a reaction against Enlightenment thinking and Modernist approaches to literature (e.g. SB Beckett, 1938-70; T Morrison, 1970-92).

1967-2004: *Columbian literature* ⁻ *Novels* and *short stories* (GG Márquez).

1970-79: *Argentinian literature* ⁻ *Poems* and *epic verses* (J Hernández).

1977-95: *US literature due to Russians* ⁻ *Poetry* and *essays* (JA Brodsky).

1985 - 2010: *German literature due to Transylvanian Saxons* ⁻ *Novels* (H Müller).

6.3. Timescale marking

Despite the stronger national traditions and the weaker international links, development of world literature and movements presented in the subchapter 6.2 took place in successive stages roughly delimited by the transitional times:

\ 3050 BC → 5065 BP \ *Pre-classical Egyptian*
\ 2700 BC → 4715 BP \ *Classical Egyptian*
\ 2300 BC → 4315 BP \ *Sumerian-Akkadian*
\ 1900 BC → 3915 BP \ *Pre-classical Sanskrit*
\ 1500 BC → 3515 BP \ *Classical Sanskrit*
\ 1100 BC → 3115 BP \ *Classical Chinese*
\ 750 BC → 2765 BP \ *Classical Greek, Persian*
\ 350 BC → 2365 BP \ *Classical Roman*
\ AD 40 → 1975 BP \ *Imperial Roman*
\ AD 430 → 1585 BP \ *Byzantine, Classical Japanese*
\ AD 820 → 1195 BP \ *Mediaeval European*
\ AD 1210 → 805 BP \ *Renaissance European*
\ AD 1600 → 415 BP \ *Worldwide*
\ AD 1990 → 25 BP \ *Modern.*

With respect to the initial time of life c.4200 million years ago, the above transitional times are reconsidered, also in million years, approximately as follows:
\ 4200 - 0.005065 = 4199.994935 \ 4200 - 0.004715 = 4199.995285 \
\ 4200 - 0.004315 = 4199.995685 \ 4200 - 0.003915 = 4199.996085 \
\ 4200 - 0.003515 = 4199.996485 \ 4200 - 0.003115 = 4199.996885 \
\ 4200 - 0.002765 = 4199.997235 \ 4200 - 0.002365 = 4199.997635 \
\ 4200 - 0.001975 = 4199.998025 \ 4200 - 0.001585 = 4199.998415 \
\ 4200 - 0.001195 = 4199.998805 \ 4200 - 0.000805 = 4199.999195 \
\ 4200 - 0.000415 = 4199.999585 \ 4200 - 0.000025 = 4199.999975 \.

More precisely, the literature emerged and developed during the latest $N/10^5 = \varepsilon$ sequences of the writing-printing process (see table at the end of subchapter 5.3), which have been delimited by values of argument \1.88058\...\1.88059\...\1.88060\...\1.88061\. Dividing these latest sequences into $N/10^6 = \zeta$ sub-sequences delimited by transitional times $t_\zeta = t_\bullet \cdot tanh(\zeta)$, the development of literature can be integrated into the evolution of life with its final time $t_\bullet = 4400$ million years from the origin of life, such as:

$$t_{1.880587} = t_\bullet \cdot tanh(1.880587) \approx 4199.994896;$$
$$t_{1.880588} = t_\bullet \cdot tanh(1.880588) \approx 4199.995286;$$
$$t_{1.880589} = t_\bullet \cdot tanh(1.880589) \approx 4199.995677;$$
$$t_{1.880590} = t_\bullet \cdot tanh(1.880590) \approx 4199.996068;$$
$$t_{1.880591} = t_\bullet \cdot tanh(1.880591) \approx 4199.996459;$$
$$t_{1.880592} = t_\bullet \cdot tanh(1.880592) \approx 4199.996850;$$
$$t_{1.880593} = t_\bullet \cdot tanh(1.880593) \approx 4199.997241;$$
$$t_{1.880594} = t_\bullet \cdot tanh(1.880594) \approx 4199.997632;$$
$$t_{1.880595} = t_\bullet \cdot tanh(1.880595) \approx 4199.998023;$$
$$t_{1.880596} = t_\bullet \cdot tanh(1.880596) \approx 4199.998414;$$
$$t_{1.880597} = t_\bullet \cdot tanh(1.880597) \approx 4199.998805;$$
$$t_{1.880598} = t_\bullet \cdot tanh(1.880598) \approx 4199.999196;$$
$$t_{1.880599} = t_\bullet \cdot tanh(1.880599) \approx 4199.999586;$$
$$t_{1.880600} = t_\bullet \cdot tanh(1.880600) \approx 4199.999977.$$

Finally, the timeline and sequences of world literature are displayed in the next table.

$N/10^6 =$ ζ	Time (million years)		Sequences in development of literature
	from origin $t_\zeta = t_\bullet \cdot tanh(\zeta)$	from present $t_\zeta - 4200$	
	4400	+200	
...	...	...	
1.880601	4200.000368	+0.000368 (AD2383)	—
1.880600	4199.999977	-0.000023 (AD1992)	_Modern
1.880599	4199.999586	-0.000414 (AD1601)	_Worldwide
1.880598	4199.999196	-0.000804 (AD1211)	_Renaissance European
1.880597	4199.998805	-0.001195 (AD820)	_Mediaeval European
1.880596	4199.998414	-0.001586 (AD429)	_Byzantine, Classic Japanese
1.880595	4199.998023	-0.001977 (AD38)	_Imperial Roman
1.880594	4199.997632	-0.002368 (353BC)	_Classical Roman
1.880593	4199.997241	-0.002759 (744BC)	_Classical Greek, Persian
1.880592	4199.996850	-0.003150 (1135BC)	_Classical Chinese
1.880591	4199.996459	-0.003541 (1526BC)	_Classical Sanscrit
1.880590	4199.996068	-0.003932 (1917BC)	_Pre-classical Sanscrit
1.880589	4199.995677	-0.004323 (2308BC)	_Sumerian-Akkadian
1.880588	4199.995286	-0.004714 (2699BC)	_Classical Egyptian
1.880587	4199.994896	0.005104 (3089BC)	_Pre-classical Egyptian
...	...	...	
0	0	-4200	

The sequences delimited in this table are succeeding on an extensive interval of time, each of them containing series of written literary works which survived over millennia by their historical and artistic qualities, altogether constituting a distinctive and invaluable heritage of the world. Literary works, styles, schools and movements were and are

174

analyzed by literary theorists and critics in order to unveil their qualities and imperfections in general, and outstanding achievements in particular.

A historical survey of the field's most important figures, schools and movements is updated annually by the *John Hopkins Guide to Literary Theory and Criticism*, published by John Hopkins University Press. This comprehensive work is compiled by 275 specialists from around the world, and includes more than 300 alphabetically arranged entries and subentries on critics and theorists, critical schools and movements, as well as critical and theoretical innovations of specific countries and historical periods.

References

3100-3000BC: <u>Narmer</u>, ancient Egyptian pharaoh, was the unifier of Egypt and founder of the First Dynasty; he became known by two *seals* from the necropolis of Abydos (later Hierakonpolis).

2600-2500BC: <u>Gilgamesh</u> (Bilgamesh), Sumerian 5th king of Uruk, built the city walls of Uruk to defend his people from external threats, and, according to the *Tummal Inscription*, he and his son rebuilt the sanctuary of the goddess Ninlil; by 2150-2000BC he became the central character in the oldest known poem, called *Epic of Gilgamesh*, later translated into Akkadian.

760-710BC: <u>Homer</u> (ancient Greek Ὅμηρος), Greek epic poet, was traditionally the author of the *Iliad* describing the Trojan War, and the *Odyssey* describing the adventures of the hero Odysseus after the fall of Troy, as the oldest European poems in dactylic hexameter.

600-553BC: <u>Zarathustra</u> (Zoroaster), Persian religious leader and prophet, apparently had visions of Ahura Mazda, which led him to preach against polytheism, and then founded *Zoroastrianism*, using the Avestan language to write the first part of *Avesta*.

498-440BC: <u>Pindar</u>, Greek lyric poet, produced many works, from them being still extant *Pythian Ode X*, odes for tyrants of Syracuse and Macedonia, as well as for free cities of Greece; *hymns* to gods, paeans, dithyrambs, mimic dancing songs, convivial songs, dirges, but unfortunately only fragments are extant; and his entirely preserved poem *Epinikia* 'Triumphal Odes'.

484-458BC: <u>Aeschylus</u>, Greek tragic dramatist, wrote about 60 plays, from which only several are extant, namely *The Persians*; *Seven against Thebes*; *Prometheus Bound*; *Suppliants*; and *Orestia*.

468-401BC: <u>Sophocles</u>, Athenian tragedian and one of great figures of Greek drama, was author of *Ajax*; *Antigone*; *Oedipus Tyrannus*; *Trachiniai* 'Women of Trachis'; *Electra*; *Philoctetes*; and *Oedipus Coloneus*.

431-406BC: <u>Euripides</u>, Greek tragic dramatist, produced *Medea*; *Andromache*; *Supplices*; *Troades*; *The Women of Troy*; *Phoenissae*; *Orestes*; *The Bacchae*; and *Iphigenia in Aulidensis*.

424-388BC: <u>Aristophanes</u>, Greek comic dramatist, wrote *Hippeis* 'Knights', *Nephelai* 'Clouds', *Sphekes* 'Wasps', *Eirene* 'Peace', *Ornithes* 'Birds', *Lysistrata* 'Destroyer of Armies', *Thesmophoriazusae* 'The Woman attending the Thesmophoria', *Batrachoi* 'Frogs', *Ecclesiazusae* 'Women in Parliament', and *Plutus*.

285-230BC: <u>Timon of Phlius</u>, Greek philosopher and poet, was author of *satyr-plays, comedies, tragedies, epic poems*, and the famous series of *Silloi* as satirical mock-heroic poems parodying and insulting most of the earlier Greek philosophers.

250-230BC: <u>Manetho</u>, Egyptian historian and priest, whose works dealt with Egyptian matters but written solely in Greek language; from his works the best known and studied is the three-volume *Aegyptiaca* 'History of Egypt', valuable for its chronology of the reigns of pharaohs; and others including *Against Herodotus*, *The Sacred Book*, *On Antiquity and Religion*, *On Festivals*, *On the Preparation of Kyphi*, and *Digest of Physics*.

166-160BC: <u>Publius Terentius</u> (Terence), Roman comic dramatist, wrote comic dramas, such as *Andria*; *Hecyra* 'Mother-in-Law'; *Heauton Timoroumenos* 'Self-Tormentor'; *Eunuchus*; *Phormio*; and *Adelphi* 'Brothers'.

80-43BC: <u>Marcus Tullius Cicero</u>, Roman orator, statesman and man-of-letters, elaborated speeches such as *Pro Roscio Amerino*; *Pro Lege Manilia*; *Pro Sestio*; *Pro Milone*; *Philippics*; as well as essays *De Senectute* 'On Old Age'; *De Amicitia* 'On Friendship'; and *De Oficiis* 'On Duty', which remained style models.

70-27BC: <u>Marcus Terentius Varro</u>, Roman scholar and writer, produced works including *Saturae Menippeae* 'Menippean Satires', *Antiquitates Rerum Humanorum et Divinarum* 'Antiquities in Matters Human and Divine', *De Lingua Latina* 'On the Latin Language', *De Re Rustica* 'Country Affairs', and the encyclopaedic *Disciplinarum Libri IX*.

37-19BC: <u>Virgil</u> (Publius Virgilius Maro), Roman poet and one of the greatest of antiquity, started with works such as *Eclogues*, pastorals; continued with *Georgics*, or *Art of Husbandry*, in four books; and culminated with his famous *Aeneid*, representing a national epic based on the story of Aeneas the Trojan, legendary founder of the Roman nation, and of the Julian family.

35-15BC: <u>Quintus Horace</u>, Roman poet and satirist, wrote remarkable books such as *Satires*; *Epodes*; three *Odes*; *Epistles*; *Carmen Seculare*; a fourth *Ode*, and three more epistles, including *Ars Poetica*, with profound influence on poetry and literary criticism in the 17th-18th centuries Europe.

22-15BC: <u>Sextus Propertius</u>, Roman elegiac poet, was author of books of *Poems* dedicated to Maecenas, the first of them being devoted to the central figure of his inspiration, the mistress Cynthia, which was the only one published during his lifetime.

15BC-AD17: <u>Ovid</u> (Publius Ovidius Naso), Roman poet, had literary success with a collection of love poems entitled *Amores* 'Loves'; followed by *Heroides* 'Heroines', *Ars Amandi* or *Ars Amatoria* 'The Art of Love', *Remedia Amoris* 'Cures for Love'; and, after was banished by Augustus and exiled to Tomis (now Constanta, Romania), he completed the great work *Metamorphoses*, a collection of mythological tales in 15 books, and also elegies, called *Tristia* 'Sorrows', and *Epistolae ex Ponto* 'Letters from Pontus Euxinus', *Ibis*, and *Halieutica* 'Fishing Matters'.

AD45-65: <u>Lucius Annaeus Seneca</u>, Roman Stoic philosopher, statesman and tragedian, wrote three treatises *Consolationes*, and then *Epistolae morales ad Lucilum*, a scathing satire entitled *Apocolocyntosis divi Claudii* 'The Pumpkinification of the Divine Claudius', philosophical *On the Shortness of Life*, and *Tenne Tragedies*.

70-95: <u>Marcus Fabius Quintilianus</u>, Roman rhetorician, became reputed for writing the great work *Institutio Oratoria* 'Education of an Orator', a complete system of rhetoric in 12 books, remarkable for its sound critical judgements, purity of taste, and perfect familiarity with literature of oratory.

80-100: Silius Italicus (Tiberius Catius Asconius Silius Italicus), Latin poet and politician, wrote the longest surviving Latin poem, entitled *Punica*, an epic in 17 books on the Second Punic War.

155-170: Lucian, Greek satirist and rhetorician, wrote in the elegant Attic style works including *Deorum Dialogi* 'Dialogues of the Gods', *Mortuorum Dialogi* 'Dialogues of the Dead', *Charon*, and *Vera Historia* 'True History'.

160-180: Lucius Apuleius, Latin-language prose writer, was author of the lewdly picaresque novel *Metamorphoses*, or *Asinus Aureus* 'The Golden Ass', a tale of adventure containing elements of magic, satire and romance; and other works such as *Apologia* 'Apology', an eloquent speech in defence.

180-220: Athenaeus of Naucratis, Greek rhetorician and grammarian, produced *Deipnosophistae* 'Banquet of the Learned', a collection of anecdotes and excerpts from ancient authors, arranged as a scholarly dinner-table conversation; from which 15 books survived.

355-363: Flavius Claudius Julian, Roman emperor, was author of *Epistles*; *Orations*; as well as satiric *Caesars*, and *Misopogon* which are still extant, but his chief work *Kata Christianon* is lost.

380-394: Gregory of Nyssa, Greek Christian theologian, produced his main works *Twelve Books against Eunomius*, a treatise on Trinity; several ascetic treatises, sermons, 23 epistles; and his great *Catechetical Oration*.

380-410: Marcus Aurelius Clemens Prudentius, Latin Christian poet, was author of poems such as *Cathemerinon Liber*; *Peristephanon*; *Apotheosis*; *Hamartigenia*; *Psychomachia*; *Contra Symmachum*; and *Diptychon*.

385-407: John Chrysostom, Archbishop of Constantinople and Early Church Father, wrote *Homilies on Jews and Judaizing Christians*, and *The Cathechetical Homily*; treatises *Against Those Who Oppose the Monastic Life*, and *On the Priesthood*; as well as *Instructions to Cathechumens*, and *Incomprehensibility of the Divine Nature*.

540-565: Procopius of Caesarea, Byzantine prominent late antique scholar, accompanied the general Belisarius against Persians, Vandals and Ostrogoths; becoming the principal historian of his century, known by his writings the *Wars of Justinian*, the *Building of Justinian*, and celebrated *Historia Arcana* 'Secret history'.

580-600: Antar ('Antarah Ibn Shaddad al-'Absi), pre-Islamic Arab poet and warrior, was author of odes and poems, which became known as one of *Seven Golden Odes* of Arabic literature, his main work being the *Mu'allaqat*, comprising the well-known poem *Romance of Antar*, as source of the 10th century model of Bedouin heroism and chivalry.

690-710: Kakinomoto no Hitomaro, Japanese poet of the late Asuka period, wrote *long poems* such as 'In the sea of ivi clothed Iwami', 'The Bay of Tsunu', and 'I loved her like the leaves'; 19 of his poems were included in the great

Man'yōshū 'Collection of a Myriad Leaves'; in Japan he is considered one of the most prominent among the Thirty-six Poetry Immortals.

720-745: <u>John of Damascus</u>, St, Byzantine Syrian monk and priest, considered 'the last of the Fathers' of the Eastern Orthodox church, became known by his *Assumption of Mary*, and then wrote works such as *Fountain of Knowledge*, also known as 'The Fountain of Wisdom', *Letter on the Thrice-Holy Hymn*, *Octoechos* (a church's service book), and *On Dragons and Ghosts*.

724-736: <u>Yamabe no Akahito</u>, poet of the Nara period in Japan, wrote *13 long poems* and *37 short poems* included in the great anthology of Japanese poetry, called *Man'yōshū*; he is regarded as one of the *kami* of poetry, and also one of the Thirty-six Poetry Immortals in his country.

948-980: <u>Egill Skallagrímsson</u>, Icelandic skaldic poet and warrior, composed the eulogy *Höfuðlausn* 'Head Ransom', the great lament *Sonatorrek* 'On the Loss of Sons', and the verse-sequence *Arinbjarnarkviða* 'The Lay of Arinbjörn'.

980-1010: <u>Firdausi</u> (Hakim Abul-Qasim Firdawsi Tusi), Persian poet, produced pieces called *kasidas* and *ghazals*; also *Yusuf u Zulaykha*, a story of Joseph and Potiphar's wife; but his masterpiece is *Shah Náma* 'Book of Kings', mainly composed of mythological and fanciful incidents, based on events from Persian annals.

1090-1114: <u>Nestor the Chronicler</u>, Ukrainian-Russian medieval writer, was author of the *Primary Chronicle*; *Life of the Venerable Theodosius of the Kiev Caves*; *Life of the Holy Passion Bearers*; *Boris and Gleb*; and the so-called *Reading*.

1137-48: <u>Anna Comnena</u>, Byzantine princess, wrote the *Alexiad*, a text-book describing the life of her father, Emperor Alexius I Comnenus, and containing a valuable account of the First Crusade.

1170-90: <u>Chrétien de Troyes</u>, French poet et trouvère, was author of poems *Érec et Énide*; *Cligès*; *Yvain, the Knight of the Leon*; and *Lancelot, the Knight of the Cart*; as well as romances such as unfinished *Perceval, ou le Conte du Graal* 'Percival, or the Story of the Holy Grail'; most of them dealing with *Arthurian* legends and subjects, and originating the character *Lancelot*.

1175-86: <u>Gligorije the Pupil</u>, early Serbian writer, produced a 362-page illuminated manuscript *Gospel Book* on parchment with rich decorations, a masterpiece of calligraphy and illustrations, which also explains the origin of Cyrillic script and its letters (National Museum of Serbia, Belgrade).

1195-1230: <u>Walther von der Vogelweide</u>, Middle High German (Austrian) lyric poet, was author of stories and poems, such as *Ich saz ûf eine steine* 'Sitting on a Stone'; collections of sayings *Reichssprüche* 'Sayings of Holy Empire', *Papstsprüche* 'Sayings about Popes', and *Kaisersprüche* 'Sayings about the Emperors'; as well as a series of songs.

1200-18: <u>Wolfram von Eschenbach</u>, German knight and poet, wrote lyrics, including *Taglieder*, love songs, short epic *Willehalm*; and most notable poem *Parzival*, with the Holy Grail's history as its main theme, which later inspired

R Wagner for the libretto of his opera 'Parsifal'.

1250-65: <u>Beatrijs van Nazareth</u> (Beatrice of Nazareth), Flemish Cistercian nun, was the first prose writer in early Dutch language, being known by her *Seven Manieren van Heilige Minnen* 'The Seven Ways of Holy Love'.

1263-84: <u>Jacob van Maerlant</u>, Flemish scholar and poet, was author of works such as *Van der Naturen Bloeme* 'The Flower of Nature', and the rhymed chronicle *De Spiegel Historiael* 'The Mirror of History'.

1307-20: <u>Dante Alighieri</u>, Italian poet, created the great *Divina Commedia* 'Divine Comedy', narrating a journey through Hell and Purgatory to Paradise, by which Italian was established as literary language, and in addition wrote *De Monarchia* (in Latin) expounding Dante's theory of divinely intended government of the world by a universal pope, and left unfinished *De Vulgari Eloquentia* discussing the origin and division of languages.

1341-70: <u>Francesco Petrarch</u>, Italian poet and scholar, wrote the epic poem *Africa*, historical prose *De Viris Illustribus*, dialogues *De Contemptu Mundi*, treatises *De Otio Religiosorum*, and *De Vita Solitaria* which influenced G Boccaccio, as well as his later *Opera Omnia*; altogether with unprecedented influence across Europe.

1350-70: <u>Giovanni Boccaccio</u>, Italian writer, produced outstanding work *Decameron* with medieval subject matter and classical form, mythological *De genealogia deorum gentilium* 'The Genealogies of the Gentile Gods', and treatises such as *De claris mulieribus* 'Famous Women', and *De Montibus* 'On Mountains'.

1360-80: <u>Hafez-e Shirazi</u>, Persian poet, became known and is remembered by his remarkable *ghazals*, and the collection *Diwan*, comprising over 570 poems.

1369-87: <u>Geoffrey Chaucer</u>, English poet, produced first his work *Book of the Duchess*, and then best known *The Canterbury Tales* by which the southern English dialect was established as the English literary language; as well as *The Parliament of Fowls*; *The House of Fame*; *Troilus and Cressida*; and *The Legend of Good Woman*.

1395-1410: <u>Andrew of Wyntoun</u>, Scottish chronicler, wrote *The Orygynale Cronykil of Scotland*, as a specimen of old Scots, covering a period from the creation until 1406 and outlining the geography and history of ancient and mediaeval Scotland.

1400-20: <u>Sophonia of Ryazan</u>, Russian priest and epic writer, was author of *Beyond the River Don*, commemorating the 1380-victory of Russians over Tatars at Kulikovo.

1430-60: <u>John Capgrave</u>, English chronicler and theologian, wrote *Nova legenda Angliae*; *De illustribus Henricis*; *Vita Humfredi Ducis Glocestriae*; and *A Chronicle of England from the Creation to 1417*.

1440-71: <u>Laurentius Valla</u>, Italian humanist, produced works including *De Donatione Constantini Magni* 'On the Donation of Constantine the Great', and *De Elegantia Latinae Linguae* 'Elegances of the Latin Language'.

1450-90: <u>Cosimo de' Medici</u> and <u>Lorenzo de' Medici</u>, Florentine statesmen and patrons of literature and art, founded the *Europe's first library*, worked on lyric poetry, and promoted *Florence as a centre of new learning*.

1463-70: <u>François Villon</u>, French poet, produced works mainly consisting of *Petit Testament*, and *Grand Testament*, with 40 and 172 eight-syllabic octaves respectively.

1469-72: <u>Afanasy Nikitin</u>, Russian merchant and writer, described his travel to India in *The Journey Beyond Three Seas*, a valuable source of information on the social system, government, economy, religion, lifestyle and natural resources of India of that time, and meanwhile unveiling his narrative style of writing.

1490-1520: <u>Marko Marulić</u>, Renaissance Croatian poet and Christian humanist, was author of the Latin tract *De institutione bene vivendi per exempla sanctorum* 'On the institution of living well according the examples of the saints', which earned him the title 'father of Croatian literature', and other writings in Latin such as *Evanglistarium*, and religious epic *Davidiad*; as well as writings in Croatian including the poems *Molitva suprotiva Turkom* 'Prayer against the Turks' and *Judita*, biblical *Suzana*, then *Carnival and Lent*, and seculary poetry *Anka the satire*.

1506-32: <u>Ludovico Ariosto</u>, Italian Renaissance poet, was author of comedies, satires, sonnets, and Latin poems, such as the great poem *Orlando Furioso* 'Orlando Enraged' developing epic tale of Roland 'Orlando'.

1512-21: <u>St Neagoe Basarab</u>, Wallachian Prince, wrote *The Teachings of Neagoe Basarab to his Son Theodosie*, with subjects including philosophy, diplomacy, morals and ethics.

1515-35: <u>Gill Vicente</u>, Portuguese dramatist, wrote plays and farces such as *Inês Pereira, Juiz da Beira*, and three *Autos das barcas* (*Inferno, Pugatório*, and *Glória*) which are his best.

1527-37: <u>Marco Girolamo Vida</u>, Italian Latin poet, produced poems on silk culture and chess, Latin orations and dialogues, as well as *Christias*, and *De Arte Poetica* 'On the Art of Poetry'.

1532-53: <u>François Rabelais</u>, French monk and Renaissance satirist, was author of books entitled *Pantagruel, Gargantua, Tiers Livre* 'The Third Book', *Quart Livre* 'Fourth Book', as well as scraps and notes for a fifth book called *L'Isle sonante*, which remain the most astonishing treasury of wit, wisdom, common sense, and satire.

1536-60: <u>Peter Zoranić</u>, Croatian Renaissance writer, produced the novel *Planine* 'Mountains', a work of prose fiction considered to be the first Croatian novel.

1558-72: <u>Luis de Camoëns</u>, Portuguese Renaissance poet, was author of the great epic poem *Os Lusiadas* 'The Lusiads' describing history of Portugal, and also plays, sonnets, and lyrics; as foundation of literary Portuguese language.

1562-93: <u>Torquato Tasso</u>, Italian Renaissance poet, wrote verses and

181

philosophical dialogues, pastoral play *Aminta*, tragedy *Il Re Torrismondo* 'King Torismondo', epic poem *Gerusalemme Liberata* 'The Recovery of Jerusalem' describing the First Crusade, and *Gerusalemme Conquistata* 'Jerusalem Conquered'.

1569-88: <u>Michel Eyquem de Montaigne</u>, French Renaissance essayist, produced the sceptical philosophical *Apologie de Raymon Sebond* 'Apologia for Raymond Sebond', and provided a major contribution to literary history introducing the new literary genre *Essais* 'Essays'.

1579-96: <u>Edmund Spenser</u>, English poet, was author of works such as *Shepheard's Calender*; *The Faerie Queene*; *Mother Hubberd's Tale*; *Colin Clout's Come Home Again*; *Complaints*; *The Early Tears of the Muses*; sonnet sequence *Amoretti*; supreme marriage poem *Epithalamion*; then *Four Hymns*, and *Prothalamion*; as well as prose *View of the Present State of Ireland*.

1580-95: <u>Thomas Lodge</u>, English dramatist, published his *Defence of Poetry*; *An Alarum against Usurers*; *The Detectable Historie of Forbonius and Priscilla*; *Scillaes Metamorphosis*; *Rosalinde*, supplying W Shakespeare with many incidents in 'As You Like It'; also a collection of poems, namely *Phillis*, and *A Fig for Momus*.

1582-93: <u>Thomas Watson</u>, English lyric poet, was author of works such as sonnets *Hecatompathia or Passionate Century of Love*, and *The Tears of Fancie*, which were probably studied by W Shakespeare.

1585-1605: <u>Miguel de Cervantes Saavedra</u>, Spanish Renaissance novelist, produced works, such as the pastoral romance *La Galatea*; the plays *La Numancia*, and *El trato de Angel*; and the immediately popular great writing *Don Quixote*; as well as tales, and a poem, founding the Spanish literary language.

1585-1606: <u>William Warner</u>, English poet, became known by works such as *Pan his Syrinx Pipe*, and a long metrical history in 14-syllable verse named *Albion's England*.

1588-93: <u>Christopher Marlowe</u>, English Renaissance dramatist, produced noticeable writings, including *Tamburlaine the Great*, a renewed style of English tragedy; *The Tragical History of Dr Faustus*, a series of detached scenes; *The Jew of Malta*, an uneven work; *Edward II*, a mature play; as well as *The Tragedy of Dido*, and *Hero and Leander*, both unfinished.

1590-1613: <u>William Shakespeare</u>, English Renaissance playwright, poet and actor, created great works, including early plays *The Two Gentlemen of Verona, Henry VI, Titus Andronicus, The Taming of the Shrew, The Comedy of Errors, Love's Labour's Lost, Romeo and Juliet*; histories *Richard III, Richard II, King John, Henry IV, Henry V*; later comedies *A Midsummer Night's Dream, The Merchant of Venice, The Merry Wives of Windsor, Much Ado About Nothing, As You Like It, Twelfth Night, Troilus and Cressida, Measure for Measure, All's Well That Ends Well*; Roman plays *Julius Caesar, Antony and Cleopatra, Coriolanus*; later tragedies *Hamlet, Othello, Timon of Athens, King Lear, Macbeth*; late plays *Pericles, The Winter's Tale, Cymbeline, The*

Tempest, Henry VIII; as well as non-dramatic works *Venus and Adonis, The Rape of Lucrece, 'The Phoenix and the Turtle'*; and finally *Sonnets*, and *'A Lover's Complaint'*.

1595-1618: George Chapman, English dramatist, produced *The Blind Beggar of Alexandria, The Gentleman Usher, Tragedie of Charle, Duke of Byron, The Widow's Tears, Caesar and Pompey, Euthymiae and Raptus, Petrarch's Seven Penitentiall Psalmes, The Divine Poem of Musaeus*, and *The Georgicks of Hesiod*.

1598-1607: John Marston, English dramatist and satirist, was author of the poem *The Metamorphosis of Pygmalion's Image and Certain Satires*; tragedies *Antonio and Mellida*, and *Antonio's Revenge*; comedy *The Malcontent*; as well as plays *The Dutch Courtesan*; *Parasitaster, or the Fawn*; *Sophonisba*; and *What You Will*.

1598-1617: John Donne, English poet, wrote passionate and erotic poems *Songs and Sonnets*; *Satires*; *Elegies*; verse *Anniversaire*; religious temper in lyrical form *Divine Poems*; prose works *Pseudo-Martyr*, and *Biothanatos*; as well as two sonnet sequences *La Corona*, and *Holy Sonnets*.

1600-24: Thomas Dekker, English dramatist and pamphleteer, was author of comedies *The Shoemaker's Holiday, or the Gentle Craft*; *The Pleasant Comedy of Old Fortunatus*; and *The Roaring Girl*; together with T Middleton, powerful drama *The Honest Whore*; with J Webster, plays *Famous History of Sir Thomas Wyat*; *Westward Ho!*; and *Northward Ho!*; and with F and W Rowley, powerful tragedy *The Witch of Edmonton*; as well as pamphlets *The Wonderful Year*, and *The Bellman of London*.

1602-34: Lope de Vega, Spanish dramatist and poet, produced works such as poem *Angelica*, miscellaneous *Rimas*, romance *Peregrino en su Patria* 'The Pilgim of Casteele', religious pastoral *Pastores de Belén*, miscellanies *Filomana* and *Circe*, and epic *Rimas de Tomé de Burguillos*.

1606-14: Benjamin Jonson, English dramatist, became known by his masterpieces *Volpone*, a satire on senile sensuality and greedy legacy hunters; *The Silent Woman*, a farcical comedy involving heartless hoax; *The Alchemist*, with a plot and strict adherence to the unities; and *Bartholomew Fair*, unveiling author's anti-Puritan prejudices.

1607-37: Thomas Heywood, English dramatist, poet and actor, wrote poetry, such as *Nine Bookes of Various History concerning Women*; plays, including domestic tragedy *A Woman Kilde with Kindnesse*, and *The English Traveller*; together with W Rowley, *Fortune by Land and Sea*; then *The Rape of Lucrece*, notable for its songs; *A Challenge for Beautie*, expressing tenderness; and *The Royal King and Loyall Subject*, stressing the doctrine of passive obedience to kingly authority.

1611-24: John Fletcher, English dramatist, produced plays *The Faithful Shepherdess*; *The Humorous Lieutenant*; and *Rule a Wife and Have a Wife*; together with F Beumont, *Philaster*, *A King and No King*; and *The Maid's Tragedy*; as well as, together with W Shakespeare, *Two Noble Kinsmen*, and

Henry VIII.

1612-23: <u>John Webster</u>, English dramatist, wrote two best known tragedies *The White Devil*, and *The Duchess of Malfi*; and also works produced in collaboration.

1613-25: <u>Luis de Góngora</u>, Spanish lyric poet, was author of long poems such as *Soledades* 'The Solitudes', *Polifemo* 'Polyphemus and Galatea', and *Piramo y Tisbe* 'Pyramus and Thisbe', written in an affected style later called 'gongorism'.

1620-44: <u>Francis Quarles</u>, English religious poet, became known for his verse, including *A Feast of Wormes*; *Argalus and Parthenia*; *Divine Poems*; *The Historie of Samson*; *Divine Fancies*; and *Emblems*; as well as prose such as *Enchyridion*, and *The Profest Royalist*.

1620-47: <u>Grigore Ureche</u>, Moldavian chronicler, was author of great *The Chronicles of the Land of Moldavia*, showing the common Romanian origin of Moldavians, Wallachians and Romanians of Transylvania.

1622-41: <u>George Wither</u>, English poet and pamphleteer, produced a book of five pastorals, *The Shepherds Hunting*; love elegy *Fidelia*; satirical *Wither's Motto*; his main poem *Fair Virtue, or the Mistress of Philarete*; then *Hymns and Songs of the Church*; *Psalms of David translated*; *Emblems*; and *Hallelujah*.

1630-40: <u>Grigore Ureche</u>, Moldavian chronicler, wrote a well-documented *The Chronicles of the land of Moldavia*, first asserting the Romance character of the Moldavian (Romanian) language.

1635-80: <u>Pedro Calderón de la Barca</u>, Spanish dramatist, produced famous dramas *La vida es sueño* 'Life's a Dream', and *El alcalde de Zalamea* 'The Mayor of Zalamea'; also writings *autos sacramentales* divided into biblical, classical, ethical, 'cloak and sword plays', and dramas of passion, the first of them being *El divino Orfeo* 'The Divine Orpheus'.

1638-74: <u>Pierre Corneille</u>, French dramatist, created the comedy *Mélite* 'Melite'; plays *L'Aveugle de Smyrne* 'The Blind Man of Smyrna', and *La Grande pastorale* 'The Great Pastoral', great *Médée* 'Medea'; the celebrated *Le Cid*; tragedies *Horace* 'Horatius', *Cinna* 'Cinna's Conspiracy', and *Polyeucte* 'Polyeuctes'; comedy *Le Menteur* 'The Mistaken Beau'; as well as *Pulchérie*, and *Suréna*.

1641-67: <u>Madeleine de Scudéry</u>, French novelist, was author of romance *Ibrahim ou l'illustre Bassa* 'Ibrahim, or The Illustrious Bassa', famous *Artamène, ou le Grand Cyrus* 'Artamenes, or the Grand Cyrus', and then *Clélie* 'Clelia', and *Mathilde d'Anguilon* 'Mathilda of Aquilar'.

1643-71: <u>John Milton</u>, English poet, became well-known by writings including *The Doctrine and Discipline of Divorce*; *Poems*; *Areopagitica*; *A Speech for the Liberty of Unlicensed Printing*; *Eikonoklastes*; two *Defensiones*; *Paradise Lost*; *Paradise Regained*; and *Samson Agonistes*.

1650-80: <u>Khushal Khan Khattak</u>, Afghan-Pashtun poet and scholar, wrote the

184

remarkable books *Baz-nama, Swat-nama, Fazl-nama, Tibb-nama, Farrukh-nama,* and *Firaq-nama.*

1653-78: <u>Izaak Walton</u>, English writer, produced celebrated *The Compleat Angler, or the Contemplative Man's Recreation,* and several biographies of contemporary personalities.

1654-59: <u>Joost van den Vondel</u>, Dutch poet and dramatist, wrote poetry and dramatic works, such as *Lucifer,* and *Jephtha,* greatly influencing German poetical revival after the devastating Thirty Years War.

1654-84: <u>Jean de La Fontaine</u>, French poet, was author of the ballad *Les Rieurs du Beau-Richard,* elegy *Pleures,* and *Nymphes de Vaux;* as well as fables, including *Le coq et la perle* 'The Cock and the Pearl', *La colombe et la fourmi* 'The Dave and the Ant', *Le lion malade et le renard* 'The Fox and the Sick Lion', *Le cheval et l'âne* 'The Horse and the Donkey', *Le vieux chat et le jeune souris* 'The Old Cat and the Young Mouse', *Le rat de ville et le rat des champs* 'The Town Mouse and the Country Mouse', *Le loup et l'agneau* 'The Wolf and the Crane', and *Le loup devenu berger* 'The Wolf who Played Shepherd'.

1658-99: <u>John Dryden</u>, English poet, wrote *Heroic Stanzas,* and *Astrea Redux;* verse plays *The Indian Emperor,* and *Aurungzebe; Annus Mirabilis, The Year of Wonders;* comedy for stage *The Rival Ladies; Essay of Dramatic Poesy; Defence of the Epilogue; Marriage à la Mode;* blanck verse *All for Love;* didactic poem *Religico Laici,* and *The Hind and the Panther;* as well as *Discourse Concerning the Origin and Progress of Satire,* and finally *Fables, Ancient and Modern.*

1659-73: <u>Jean Baptiste Poquelin Molière</u>, French playwright, created comic works *Les Précieuses ridicules* 'The Conceited Young Ladies', masterpieces *Tartuffe; Le Misanthrope* 'The Misanthrope'; *Amphitryon; Le Bourgeois gentilhomme* 'The Citizen turned Gentleman'; and *George Dandin;* then *L'Avare* 'The Miser', *Les Fourberies de Scapin* 'The Cheats of Scapin', and *Malade imaginaire* 'The Imaginary Invalid'.

1660-77: <u>Nicolas Boileau</u>, French poet and critic, wrote works including *L'Art poétique* 'The Art of Poetry', comic *Lutrin* 'Lectern', and *Dialogue des héros de roman* 'A Conversation between the Heroes of Novels'.

1664-91: <u>Jean Racine</u>, French dramatist and poet, was author of valuable plays including *La Thébaïde ou Les Frères ennemis* 'The Fatal Legacy', *Alexandre le grand* 'Alexander the Great', then *Andromaque* 'Andromache', *Les Plaideurs* 'The Litigants', *Britannicus, Bérénice* 'Titus and Berenice', *Bajazet* 'The Sultaness', *Mithridate* 'Mithridates', masterpiece *Iphigénie* 'Achilles, or Iphigenia in Aulis', marvellous *Phèdre* 'Phaedre and Hippolytys', as well as plays *Ester* and *Athalie.*

1671-77: <u>William Wycherley</u>, English dramatist, produced comedies *Love in a Wood, or St James's Park,* and *The Gentleman Dancing-master,* as well as plays *The Country Wife,* and *The Plain Dealer.*

1672-87: <u>Miron Costin</u>, Moldavian political figure and chronicler, continued the work of G Ureche, writing *The Chronicles of Moldova From the Prince Aron's Rule* (to 1660), and compiled *Polish Verse History of Moldavia and Wallachia*.

1673-90: <u>Dimitrie Barila</u> (St Dosoftei, or Dositheiu), Moldavian metropolitan, scholar and poet, published the first *volume of poetry in Romanian*, a verse translation of the *Psalms*.

1675-1711: <u>Nicolae Costin</u>, Moldavian statesman and chronicler, became known by his historical writings including *The Chronicle of the Principality of Moldavia from the World's Building to 1601*, and *The Chronicle of Nicolae Mavrocordat's Rule*.

1696-1726: <u>John Vanbrugh</u>, English playwright, wrote *The Relapse*, *The Provok'd Wife*, *The Confederacy*, and unfinished *The Provok'd Husband*.

1700-08; <u>Antim Ivireanul</u>, Georgian-born Wallachian metropolitan, author and typographist, published the popular book *The Flower of the Gifts*, and *The Sermons*.

1700-16: <u>Constantin Cantacuzino</u>, Wallachian nobleman, was author of *The Political and Geographical History of the Romanian Countries*.

1704-58: <u>Jonathan Swift</u>, Anglo-Irish satirist and clergyman, was author of satires *A Tale of a Tub*, and *Gulliver's Travels*; verses *Journal to Stella*, *The Grand Question Debated*, and *Verses on His Own Death*; pamphlet *On the Conduct of the Allies*; poem *Cadenus and Vanessa*; verse satire *On Poetry; A Rhapsody*; ironical writings *Directions to Servants*, and *A Complete Collection of Genteel and Ingenious Conversation*; as well as *History of the Four Last Years of the Queen*.

1718-78: <u>François Marie Arouet Voltaire</u>, French enlightenment writer and historian, created great works, such as dramas *Œdipe* 'Oedipus', *Mahomet* 'Mohamet the Imposter', *Princesse de Navarre*, and *Irène*; poetry *La Ligue ou Henri le Grand* 'Henriade', and *Poème sur le désastre de Lisbonne* 'Poem on the Lisbon Earthquake'; philosophical tales *Lettres écrites de Londres sur les Anglais* 'Letters Concerning the English Nation', and *Candide* 'Candid'; as well as philosophical and historical writings *Traité de métaphysique* 'Treatise on Metaphysics', *Siècle de Louis Quatorze* 'The Age of Louis XIV', *Les Mœurs et l'esprit des nations* 'The General History and State of Europe', and *Dictionnaire philosophique portatif* 'The Philosophical Dictionary for the Pocket'.

1719-28: <u>Daniel Defoe</u>, English writer and adventurer, wrote books *Robinson Crusoe*; *Journal of the Plague Year*; *Moll Flanders*; *Roxana*; as well as 3-volume travel book including *Tour through the Whole Island of Great Britain*; *The Great Law of Subordination Considered*; and *Augusta Triumphans, or the Way to make London the Most Flourishing City in the Universe*.

1719-43: <u>Ion Neculce</u>, Moldavian chronicler, was author of the great work entitled *The Chronicles of the Land of Moldavia (from the Prince Dabija's Rule to the second rule of Constantin Mavrocordat)*.

1719-45: <u>Edward Young</u>, English poet, produced tragedies *Busiris*, *The*

Revenge, and *The Brothers*; satire *The Love of Fame, the Universal Passion*; poem *The Instalment*; and *The Complaint, or Night Thoughts on Life, Death and Immortality*.

1735-52: <u>Vasily Kirillovich Trediakovsky</u>, Russian poet, essayist and playwright, was author of *A new and brief way for composing of Russian verses*, a theoretical work; *A Conversation on Orthography*, a study of phonetic structure of the Russian language; and *On Ancient, Middle, and New Russian Poetry*, advocating a poetical reform.

1735-55: <u>Khana Qubadi</u>, Kurdish poet, wrote a series of poems, but his main remained work is *Şîrîn û Xesrew* 'Shirin and Khasraw'.

1740-54: <u>Samuel Richardson</u>, English novelist, wrote novels including *Pamela*; 7-volume *Clarissa, Or the History of a Young Lady*; and *Sir Charles Grandison*.

1741-51: <u>Henry Fielding</u>, English novelist, became known for his *An Apology for the Life of Mrs Shamela Andrews*; *The Adventures of Joseph Andrews and his Friend, Mr Abraham Adams*; and three-volume *Miscellanies*; *The History of Tom Jones, a Foundling*; and *Amelia*.

1742-55: <u>Prosper Jolyot de Crébillon</u>, French novelist, produced tragedies *Idoménée, Atrée et Thyeste, Électre*, and masterpiece *Rhadamiste et Zénobie*, as well as *Catalina*.

1749-81: <u>Samuel Johnson</u> (Dr Johnson), English writer, critic and lexicographer, wrote the didactic poem *The Vanity of Human Wishes*; *Dictionary of the English Language*; moral fable *Rasselas: The Prince of Abyssinia*; and monumental 10-volume *Lives of the Most Eminent English Poets*.

1753-87: <u>Carlo Goldoni</u>, Italian dramatist, was author of plays such as *La Locandiera* 'The Mistress of the Inn', *I rusteghi* 'The Boors', and *Le baruffe chiozzote* 'The Squabbles of Chioggia'; and finally his *Mémoires*.

1759-73: <u>Oliver Goldsmith</u>, Irish playwright, novelist and poet, produced *Enquiry into the Present State of Polite Learning in Europe*; novel *The Vicar of Wakefield*; poetry *The Deserted Village*; and play *She Stoops to Conquer*.

1762-73: <u>Ivan Semyonovich Barkov</u>, Russian historic writer and comic-erotic poet, became known by writing *A Brief History of Russia*, and also by his erotically works *Luka Mudischev*, and *Shameful Odes*.

1767-84: <u>Pierre Augustin Caron de Beaumarchais</u> French playwright, produced plays *Eugénie* 'The School for Rakes', and *Les Deux Amis* 'The Two Friends'; reputed satire *Mémoires du Sieur Beaumarchais par lui-même* 'Autobiography'; and famous comedies *Le Barbier de Séville* 'The Barber of Seville', and *La Folle journée ou le mariage de Figaro* 'The Follies of a Day, or The Marriage of Figaro'.

1767-1832: <u>Johann Wolfgang von Goethe</u>, German poet and dramatist, created great plays *Die Laune des Verliebten* 'The Beloved's Whim', and *Die Mitschuldigen* 'The Accomplices'; masterpiece drama *Götz von Berlichingen*;

novels *Die Lieden des jungen Werthers* 'The Sorrows of Young Werther', and *Wilhelm Meisters Theatralische Sendung* 'Wilhelm Meister's Theatrical Mission'; verse dramas *Iphigenie auf Tauris, Egmont,* and *Torquato Tasso*; epic idyll *Hermann und Dorothea*; drama *Die natürliche Tochter* 'The Natural Daughter'; and novel *Die Wahlverwandtschaften* 'The Elective Affinities'.

1768-90: <u>Carl Michael Bellman</u>, Swedish poet and songwriter, produced collections of poems set to music *Fredmans sånger* ' Fredman's songs', and *Fredmans epistlar* 'Fredman's epistles', together consisting of about 70 songs; other works such as *Blåsen nu alla!* 'All blow now'; *Ulla Winblad*; and the minuet melody *Ach du min Moder* 'Alas, thou my mother'.

1771-74: <u>Robert Fergusson</u>, Scottish poet, wrote poems in Scots and English, including *The Daft Days*; famous *Auld Reekie*; *Elegy on the Death of Scots Music*; *Hallow Fair*; *To the Tron Kirk Bell*; *Leith Races*; and satirical *The Rising of the Session*.

1774-1810: <u>Dositej Obradović</u>, Serbian-Romanian philosopher and linguist, produced works such as *Hristoitija, Etika, Venac, Damon,* and also short stories.

1778-1800: <u>Johann Gottfried Herder</u>, German critic and poet, wrote collection of folksongs *Stimmen der Völker in Liedern* 'Voices of the Peoples in Songs', treatise *Vom Geist der Ebraïschen Poesie* 'The Spirit of Hebrew Poetry', a version of *Cid*, and especially *Ideen zur Geschichte der Menschheit* 'Outlines of a Philosophy of the History of Man'.

1781-1804: <u>Johann Christoph Friedrich Schiller</u>, German dramatist and poet, created plays *Die Räuber* 'The Robbers', *Fiesko* 'Fiesco', *Kabale und Liebe* 'Cabal and Love', and *Don Carlos*; stories *Verbrecher aus verlorener Ehre* 'The Dishonoured Irreclaimable', and *Der Geisterseher* 'The Gost-Seer'; poems *An die Freude* 'Ode to Joy', and *Die Künstler* 'The Artists'; famous *Über naïve und sentimentalische Dichtung* 'On Simple and Sentimental Poetry'; celebrated *Xenien* 'Epigrams'; ballads *Der Taucher* 'The Diver', *Der Ring des Polykrates* 'The Ring of Polykrates', *Die Kraniche des Ibykus* 'The Cranes of Ibycus', and especially *Das Lied von der Glocke* 'Song of the Bell'; as well as the dramatic trilogy *Wallenstein*, the greatest historical drama in German language; then psychological study *Maria Stuart*, and half-legend *Wilhelm Tell*.

1786-96: <u>Robert Burns</u>, Scottish poet and songwriter, was author of *Poems, Chiefly in the Scottish Dialect*; and songs such as *John Anderson My Jo*; *Ae Fond Kiss*; *Ye Jacobites by Name*; *The Banks o'Doon*; *Afton Water*; *A Red Red Rose*; and *Auld Lang Syne*.

1789-1817: <u>Madame de Staël</u>, French writer, wrote the celebrated *Lettres sur Rousseau*; *Réflexions sur la paix intérieur* 'Reflexions on Civil Peace'; *Influence des passions*; *Littérature et ses rapports avec les institutions sociales* 'The Influence of Literature upon Society'; then the novel *Delphine*, and romance *Corinne*; as well as *De l'Allemagne* 'Germany'.

1792-1826: <u>Nikolay Mikhailovich Karamzin</u>, Russian poet, historian and critic, started with writing the pieces *Poor Liza*, and *Natalia the Boyar's Daughter*, continued with poems such as *Nightingale*, and *Hymn to the Fools*, and finished with the 12-volume *History of the Russian State*.

1795-1802: Friedrich Hölderlin, German poet, wrote poems, philosophical novel *Hyperion*, elegy *Menon's Laments for Diotima*, and works *Brot und Wein* 'Bread and Wine', and *Der Rhein* 'The Rhine'.

1795-1847: Robert Southey, English poet and writer, became known by his volume *Poems*; epic poem *Joan of Arc*; biographies of *Nelson*, *Wesley*, and *Bunyan*; as well as *A Vision of Judgement*; *Naval History*; and *The Doctor*.

1798-1840: William Wordsworth, English poet, wrote *Lyrical Ballads*; poems *Michael*; *Ruth*; *Lucy*; *The Solitary Reaper*; *The Recluse*; *The Prelude*; especially *The Excursion*; *Ecclesiastical Sonnets*; and *Memorials*.

1800-11: Heinrich von Kleist, German dramatist and poet, was author of the fine tale *Michael Kohlhaas*, and popular play *Prinz Friedrich von Homburg* 'The Prince of Homburg'.

1801-47: Adam Gottlob Oehlenschläger, Danish poet and playwright, produced the dramatic sketch *April the Second 1801*, piece of fantasy *Aladin*; masterpieces *Baldur hin Gode*, *Palnatoke*, and *Axel og Valborg*; tragedies *Correggio*, *Stærkodder*, *Hagbarth og Signe*, and *Erik og Abel*; verse-romances *Helge*, epic *Hrolf Krake*, and volumes such as *Kiartan og Gudrun*.

1801-48: René de Chateaubriand, French writer and statesman, was author of works such as *Atala*; *Le Génie du Christianisme* 'The Beauties of Chriastanity'; *Itinéraire de Paris à Jerusalem* 'Travels in Greece, Palestine, Egypt and Barbary'; celebrated *Mémoires d'outre-tombe* 'Memoirs from Beyond the Grave'; and autobiographical *Memoires*.

1802-27: Walter Scott, Scottish novelist and poet, produced *The Minstrelsy of the Scottish Border*; *Waverley*; *Ivanhoe*; *The Talisman*; *Rob Roy*; *Life of Napoleon*; and *Tales of a Grandfather*.

1807-27: Thomas Moore, Irish poet and composer, wrote *Irish Melodies*, *The Twopenny Postbag*, *Lalla Rookh*, *The Loves of the Angels*, and the novel *The Epicurean*.

1807-38: Franz Seraphicus Grillparzer, Austrian writer, was author of tragedy in iambics *Blanca von Castilien*, play *The Ancestress*, drama *Sappho*, trilogy *Das goldene Vlies* 'The Golden Fleece'; historical tragedies *König Ottokars Glück und Ende* 'King Ottokar's Fortune and End', and *Ein treuer Diener seines Herrn* 'A faithful Servant of his Lord'; dramas *Des Meeres und der Liebe Wellen* 'Waves of the Sea and of Love', and *Der Traum, ein Leben* 'The Dream, a Life'; and comedy *Weh dem, der lügt* 'Woe to him who lies'.

1808-37: Washington Irving, US writer, pioneered American literature, writing satirical essays *Salmagundi*, boisterous work *A History of New York, by Diedrich Knickerbocker*, miscellany of tales *The Sketch Book* and *Tales of a Traveller*, studies *The History of the Life and Voyages of Christopher Columbus* and *The Conquest of Granada*, as well as *A Tour on the Prairie* and

189

The Adventures of Captain Bonneville.

1809-24: <u>George Gordon Noel Byron</u>, English poet of Scottish antecedents, produced Popian satire *English Bards and Scotch Reviewers*, *Childe Harold's Pilgrimage*; satires and poems, including *English Bards and Scotch Reviewers*; *Childe Harold's Pilgrimage*; as well as oriental pieces *Giaour*; *Lara*; and *Siege of Corinth*; his best works being *Beppo*, *A Vision of Judgement*, and satirical *Don Juan*.

1811-17: <u>Jane Austen</u>, English novelist, wrote novels and literary works, such as *Sense and Sensibility*; *Pride and Prejudice*; *Mansfield Park*; *Emma*; and *Persuasion*.

1812-37: <u>Jacob Ludwig Carl Grimm</u> and <u>Wilhelm Carl Grimm</u>, German writers and linguists, apart of their great linguistic works, they were authors of *Kinder- und Hausmärchen* 'Children and House Tales' as a foundation of the science of comparative folklore; *Deutsche Sagen* 'German Sagas'; and *Geschichte der deutsche Sprache* 'History of German Language'.

1812-54: <u>Heinrich Heine</u>, German poet and essayist, wrote poetry, prose, and other works, including *Gedichte* 'Poems', *Lyrisches Intermezzo* 'Lyrical Intermezzo', volumes of prose *Reisebilder* 'Pictures of Travel', the well-known *Das Buch der Lieder* 'Book of Songs'; as well as *Die Romantische Schule* 'The Romantic School', *Neue Gedichte* 'New Poems', three-volume *Vermischte Schriften* 'Various Writings', and *Französische Zustände* 'French Affairs'.

1813-20: <u>Nguyen Du</u>, Vietnamese poet, published his epic poem *The Tale of Kieu*, and other works including *Bac Hanh Tap Luc* 'Travels to the North', *Nam Trung Tap Ngam* 'Various Poems', and *Thanh Hien thi tap* 'Poems of Thanh Hien'.

1813-22: <u>Percy Bysshe Shelley</u>, English lyric poet and writer, created great poetry and prose pieces, including *Queen Mob*; unfinished novella *The Assassins*; *The Revolt of Islam*; the masterpiece *Prometheus Unbounded*; intimate *Letter to Maria Gisborne*, and *The Witch of Atlas*; prose pieces *A Philosophical View of Reform*, *Essay on the Devil*, *The Defence of Poetry*; as well as *Swellfoot the Tyrant*, *Adonais*, *Epipsychidion*; and verse drama *Hellas*.

1814-22: <u>Ernst Theodor Wilhelm Hoffmann</u>, German writer, music critic and caricaturist, produced short tales included in collections *Fantasiestücke* 'Fantasies', *Nachtstücke* 'Night-time Tales', and *Die Serapionsbrüder* 'The Serapion Brothers'; also long tales such as *Elixiere des Teufels* 'The Devil's Elixirs'; and partly autobiographical *Lebensansichten des Katers Murr* 'Opinions of the Tomcat Murr'.

1815-30: <u>Fyodor Nikolaevich Glinka</u>, Russian poet and playwright, was author of the eight-volume *Letters of a Russian Officer*, and of the descriptive poem *Karelia*.

1816-20: <u>John Keats</u>, English poet, wrote splendid romances and poems such as *Hymn to Pan* and *Bacchic procession*, *The Examiner*; *Edition*; mythological poem *Endymion*; and also the volume *Lamia and Other Poems*, a landmark in English poetry, which includes romantic 'Isabella or The Pot of Basil', epical 'Hyperion', two splendid romances 'The Eve of St Agnes' and 'Lamia', great odes 'On a Grecian Urn', 'To a Nightingale', 'To Autumn', 'On Melancholy', and 'To Psyche'.

1818-50: <u>Adam Bernard Mickiewicz</u>, Polish national poet, dramatist and essayist, was author of poems *Zima miejska* 'City Winter', *Kartofla* 'Potato'; drama *Dziady* 'Forefathers' Eve'; *Crimean Sonnets*; *Konrad Wallenrod*; *Grażyna*; *Death of the Colonel*; epic poem *Pan Tadeusz*; *Lausanne Lyrics*; and dramas *Les Confederes de Bar* 'The Bar Confederates', and *Jacques Jasiński, ou les deux Polognes* 'Jacque Jasiński, or the Two Polands'.

1820-30: <u>Alexander Sergeyevich Pushkin</u>, Russian great poet and writer, created noticeable poems such as *Ruslan and Lyudmilla*, *The Prisoner of the Caucasus*, *Fountain of Bakhchisarai*, and *Tzigani*, also novel in verse *Eugene Onegin*, and blank verse drama *Boris Godunov*.

1820-51: <u>Alphonse Marie Louis de Lamartine</u>, French poet and historian, wrote the poems *Méditations*, *Harmonies poétiques et religieuses* 'Political and Religious Harmonies', *Souvenirs d'Orient* 'Recollections of a Pilgrimage to the Holy Land', *Jocelyn*, *La Chute d'un ange* 'The Fall of an Angel', *Histoire des Girondins* 'History of the Girondins', *Confidences* 'Memoirs of My Youth', *Raphaël*, *Geneviève*, *Histoire de la Restauration*, and *Le Tailleur de pierres de St-Point* 'The Stonecutter of St Point'.

1820-53: <u>Almeida Garrett</u>, Portuguese romantic poet, playwright and novelist, published his works including the poem *O Roubo das Sabinas* 'The Theft of the Sabines', classical tragedy *Catão* 'Cato', poem *Camões* 'Camoens', masterpiece play *Frei Luís da Sousa* 'Brother Luíz de Sousa'; fiction works *Viagens na Minha Terra* 'Travels in my Homeland', and *O Arco de Sant'Ana II* 'The Arch of Sant'Ana II'; as well as poem *Folhas Caídas* 'Fallen Leaves'.

1821-45: <u>James Fenimore Cooper</u>, US novelist, produced stories of sea and of Native American Indians, including *The Spy*, *The Pilot*, *The Last of the Mohicans*, *The Prairie*, *The Red Rover*, *The Bravo*, *The Pathfinder*, *The Deerslayer*, *The Two Admirals*, *Wing-and-Wing* and *Satanstoe*.

1821-55: <u>Sunthorn Phu</u>, Thailand's best known royal poet, wrote canonical works including the collections of poems *Nirat Phukaothong*, and *Nirat Suphan*, as well as saga *Phra Aphai Mani*.

1822-47: <u>Dionysios Solomos</u>, Greek poet, was author of the national poem *Hymn to Liberty*; then *Ode to the death of Lord Byron*; works *The Woman of Zakynthos*, and *The Dialogue*; poems *Lambros*, *O Kritikos* 'The Cretan'; as well as *The Free Besieged*, *O Porfyras*, and *The Whale*.

1822-82: <u>Victor Marie Hugo</u>, French poet and author, produced works including poems (*Odes and ballades*, *Les Orientales*, *Les Feuilles d'automne*, *Chants du crepuscule*, *Les Voix intérieures*, *Les Rayons et les ombres*, *Le Châtiments*, *Les Contemplations*, *Légende des siècles*, *Les Chansons des rues et des bois*); dramas (*Cromwell*, *Hernani*, *Marion Delorme*, *Le Roi s'amuse*,

Lucrèce Borgia, Marie Tudor, laude Gueux, Ruy Blas); novels (*Notre Dame de Paris, Les Misérables, Les Travailleurs de la mer, Quatre-vingt-treize*), as well as criticism and other writings (*Littérature et philosophie mêlées, Napoléon le petit, William Shakespeare, L'Homme qui rit, L'Histoire d'un crime*).

1827-45: <u>Edgar Allan Poe</u>, US poet and short-story writer, was author of the volumes of verse *Tamerlane and other Poems, Berenice*, and *Al Aaraaf*; then *The Narrative of Arthur Gordon Pym*; *Tales of the Grotesque and Arabesque*; and *The Raven and Other Poems*, thus pioneering the modern detective story.

1828-63: <u>Nathaniel Hawthorne</u>, US Novelist and short story writer, published his first novel *Fanshawe*, then short stories collected as *Twice-told Tales*, stories for children, sketches and studies collected as *Mosses from an Old Manse*, best-known *The Scarlet Letter, The House of the Seven Gables, Wonder Book, The Snow Image, The Blithedale Romance, Tanglewood Tales, The Marble Faun*, and papers collected as *Our Old Home*.

1829-33: <u>Honoré de Balzac</u>, French novelist, published novels unveiling a complete picture of modern civilization in his *Comédie humaine*, and other known works such as *Le Père Goriot, Les Illusions perdues, Eugénie Grandet*, and *Contes drolatiques*.

1829-44: <u>Henrik Arnold Thaulow Wergeland</u>, Norwegian poet and playwright, was author of the volume of lyrical and patriotic poems *Digte, første Ring* 'Poems, first circle'; then *Campbellerne* 'The Campbells', *Jan van Huysums Blomsterstykke* 'Flower-piece by Jan van Huysum', *Svalen* 'The Swallow', *Jøden* 'The Jew', *Jødinden* 'The Jewess', *Den Engelske Lods* 'The English Pilot', and *Venetianerne* 'The Venetians'.

1830-39: <u>Henry Marie Beyle Stendhal</u>, French novelist, wrote masterpieces *Le Rouge et le noir* 'Red and Black', and *La Chartreuse of Parma* 'The Charterhouse of Parma'.

1830-57: <u>Francisco Balagtas</u>, Filipino poet, became well-known by his epic *Florante at Laura*, poem *Tagalog*, and also by comedies such as *Don Nuño at Selinda, Auredato at Astrome, Clara Belmore*, and *Alamansor at Rosalinda*.

1830-69: <u>Théophile Gautier</u>, French poet and novelist, wrote poems including *Albertus*; *Comédie de la mort* 'The Comedy of Death'; as well as collection *Émaux et camées* 'Enamels and Cameos', novel *Mademoiselle de Maupin*; and informal autobiography *Ménagerie intime*.

1831-42: <u>Nikolai Vasilevich Gogol</u>, Russian novelist and dramatist, published his major work *Vechera na khutore bliz Dikanki* 'Evenings on a Farm near Dikanka'; short-story collections *Mirgorod*, and *Arabesques*; play *Revizor* 'The Inspector-General'; and great novel *Myortvye dushi* 'Dead Souls'.

1832-36: <u>Karel Hynek Mácha</u>, Czech romantic poet, was author of writings such as *Diary of Travel to Italy*, autobiographical sketches *Pictures from My Life*, and lyrical poem *Máj* 'May'.

1832-61: <u>George Sand</u>, French novelist, published her works including

Indiana, Valentine, Lélia, Jacques, La Comtesse de Rudolstadt, Mauprat, Les Beaux messieurs de Bois-Doré 'The Gallant Lords of Bois-Doré', and *Le Marquis de Villemer* 'The Marquis of Villemer'.

1833-51: <u>Petar II Petrović-Njegoš</u>, Prince-Bishop of Montenegro, Serbian Orthodox Metropolitan, philosopher and poet, was author of *The Voice of Mountaineers*; *The Cure for the Turkish Fury*; *The Serbian Mirror*; *The Ray of the Microcosm*; masterpiece *The Mountain Wreath*; and *The False Tsar Stephen the Little*.

1834-70: <u>Harriet Beecher Stowe</u>, US novelist, wrote short-story *A New England Sketch*; anti-slavery novels *Uncle Tom's Cabin*, and *Dred: A Tale of the Dismal Swamp*; works *The Minister's Wooing*, and *Old Town Folks*; as well as *Lady Byron Vindicated*.

1835-55: <u>Hans Christian Andersen</u>, Danish writer, published great stories, poetry, travel books, novels, plays, and pamphlets of fairy tales, such as *The Tin Soldier, The Emperor's New Clothes*; *The Tinderbox, The Snow Queen*; also *The Little Mermaid*, and many pamphlets of fairy tales *The Ugly Duckling*; all of them spreading across Europe.

1836-50: <u>Alexandre Dumas</u> (Dumas père), French novelist and playwright, produced novels and plays including *Isabelle de Bavière*; *Pauline*; *Acté*; the well-known *Le Comte de Monte Cristo* 'The Count of Monte Cristo', *Les Trois mousquetaires* 'The Three Musketeers', *Vingt ans après* 'Twenty Years After', *Dix Ans plus tard ou le Vicomte de Bragelonne* 'Ten Years Later or The Vicomte de Bragelonne', *La Reine Margot* 'Queen Margot', *La Dame de Monsoreau* 'The Lady of Monsoreau', *Le Collier de la Reine* 'The Queen's Necklace', and *La Tulipe noire* 'The Black Tulip'.

1837-41: <u>Mikhail Yuriyevich Lermontov</u>, Russian poet, wrote the impressive novel *Geroy nashevo vremeni* 'A Hero of Our Times', romantic verse play *Maskarad* 'Masquerade', and sensitive poetry inspired by the scenery of Caucasus.

1837-64: <u>Charles John Huffman Dickens</u>, English novelist, was author of great works including *Oliver Twist*; *Nicholas Nickleby, Barnaby Rudge, The Old Curiosity Shop, David Copperfield, A Tale of Two Cities, Great Expectations, Hard Times, Little Dorrit*, and *Our Mutual Friend*.

1838-60: <u>Eugène Labiche</u>, French playwright, wrote the novel *La Clef des champs* 'Key of the Fields', over 150 skilfully observed and crafted comedies, farces and vaudevilles, as well as *Frisette*, and *Le Voyage de M. Perrichon*.

1840-45: <u>Taras Hryhorovych Shevchenko</u>, Ukrainian poet, writer and folklorist, produced the collection of poetry *Kobzar*; *The Gypsy Fortune Teller*; tragedy *Mykyta Haidai*; and finally the poem *Zapovit* 'Testament'.

1840-60: <u>William Makepeace Thackeray</u>, English novelist, wrote pieces such as *The Paris Sketchbook*, and *Punch*; great novels *Vanity Fair*; *Pendennis*; three-volume *Henry Edmond*; *The Newcomes*; as well as *The English Humorists of the 18th century*, and *The Four Georges*.

1842-48: <u>Sándor Petöfi</u>, Hungarian poet, composed poems and songs including *A borozó* 'The Wine Drinker', volume of poetry *Versek*, the popular *János vitéz* 'Janos the Hero', poem *A régi, jó Gvadányi* 'The Good Old Gvadányi', *Cipruslombok Etelke sírjára* 'Branches of Cypress for Etelke's Tomb', novel *A hóhér kötele* 'The Hangman's Rope', the national hymn *Nemzeti dal* 'National Song', as well as the novel *A hóhér kötele* 'The Hangman's Rope'.

1845-47: <u>Emily Jane Brontë</u>, English writer, was author of verse *Gondal*, and especially powerful novel *Wuthering Heights*, a tale of love and revenge in style of the Greek tragedy.

1846-57: <u>Herman Melville</u>, US novelist, wrote short-story and poetry, starting with *Typee*, *Omoo*, and *White Jacket*, continuing with masterpiece *Moby-Dick*, and novel *Pierre*, and finishing with collection of short stories *The Piazza Tales*, and *The Confidence Man*.

1847-54: <u>Charlotte Brontë</u>, English literary figure, produced the masterpiece *Jane Eyre*; novels *The Professor*, and *Shirley*; memoires *Villette*; novel-fragment *Emma*; and stories *The Secret*, and *Lily Hart*.

1847-56: <u>Antônio Gançalves Dias</u>, Brazilian romantic poet, playwright and linguist, wrote *Primeiros Cantos* 'First Chants'; the poem *Canção do exílio*; then *Segundos Cantos* 'Second Chants', and *Últimos Cantos* 'Last Chants'; short epic poem *I-Juca-Pirama*; and unfinished *Os Timbiras* 'The Timbiras'.

1848-70: <u>Johan Ludvig Runeberg</u>, Finnish poet, was author of the famous *Fänrik Ståls sägner* 'The Tales of Ensign Stål', and also of the poem *Vårt land* 'Our land', and *The Cloud's Brother*.

1849-56: <u>Fernán Caballero</u>, Spanish novelist, wrote about 50 romances, including *La Gaviota* 'The Seagull', *Clemencia*, *Un servilón y un liberalito* 'A Groveller and a Little Liberal', and *La Familia de Alvareda*.

1850-70: <u>Ivan Sergeyevich Turgenev</u>, Russian novelist, was author of plays and novels, including the play *A Month in the Country*; study *Zapiski okhotnika* 'A Sportsman's Sketches'; novels *Ottsy I dety* 'Fathers and Sons', *Dym* 'Smoke', and *Nov'* 'Virgin Soil'; as well as powerful piece *Stepnoy Korol'Lir* 'A Lear of the Stepes'.

1850-89: <u>Alfred Tennyson</u>, English poet, produced the elegiac poem *In Memoriam*; verse novelettes *Maud: A Monodrama*, *Enoch Arden*, and *Locksley Hall Sixty Years After*; sequence of poems *Idylls of the King*; also the play *Becket*, and lyric poem *Crossing the Bar*.

1850-99: <u>Henrik Ibsen</u>, Norwegian dramatist, was author of *Samfundets støtter* 'Pilars of Society', *Et dukkehjem* 'A Doll's House', *Gengangere* 'Ghosts', *En folkefiende* 'An Enemy of the People', *Vildanden* 'The Wild Duck', *Rosmersholm*, *Fruen fra havet* 'The Lady from the Sea', *Hedda Gabler*, *Bygmester Solness* 'The Master Builder', *John Gabriel Borkman*, and *Naar vi døde vaagner* 'When We Dead Awaken'.

1851-99: <u>Leo Nikolayevich Tolstoy</u>, Russian writer, aesthetic philosopher, moralist and mystic, produced well-known works including *Istoria vcherashchnevo dnya* 'An Account of Yeasterday'; autobiographical trilogy *Detstvo* 'Childhood', *Ostrochestvo* 'Boyhood', and *Yunost* 'Youth'; sketches of Crimean War *Tales of Army Life*, and *Sevastopolskiye rasskazy* 'Sebastopol'; first great *Voinya i mir* 'War and Peace'; second great *Anna Karenina*; then the religious *V chom moya vera?* 'My Religion'; followed by *Tsarstvo Bozhye vnutri vas* 'The Kingdom of God is within You', *Khosyain i rabotnik* 'Master and Man', *Plody prosveshcheniya* 'The Fruits of Enlightenment', and *Voskreseniye* 'Ressurection'.

1855-91: <u>Walter Whitman</u>, US poet, published poems *Leaves of Grass*, *Drum Taps*, and *Sequel to Drum Taps*; prose *Specimen Days and Collect*; as well as collection of poems and prose *Good-Bye*, and *My Fancy*.

1857-69: <u>Charles Pierre Baudelaire</u>, French Symbolist poet, wrote collection of poems *Les Fleurs du mal* 'The Flowers of Evil', *Les Paradis artificiels* 'The Artificial Paradises', and *Petit Poèmes en prose* 'Little Poems in Prose', unveiling a sort of satanism, combined with the macabre, perverted and horrid.

1857-80: <u>Gustave Flaubert</u>, French novelist, published works including *Madame de Bovary*, *Salammbô*, *L'Éducation sentimentale*, *La Tentation de St Antoine*, *Trois contes (Un Cœur simple*, *La Légende de Saint Julien l'Hospitalier*, and *Hérodias)*, *Bouvard et Pécuchet*, and correspondence with G Sand.

1858-78: <u>Nikolai Alekseyevich Nekrasov</u>, Russian lyrical poet, wrote poems depicting social wrongs of peasantry, such as the unfinished narrative epic *Komu na Rusi zhit khorosho?* 'Who Can Be Happy and Free in Russia?'.

1860-81: <u>Prosper Mérimée</u>, French novelist, was author of novels such as *Carmen*, *La Vénus d'Ille*, *Arsène Guillot*, and *L'Abbé Aubain*; as well as letters called *Lettres à une inconnue*, and *Lettres à une autre inconnue*.

1860-87: <u>Edmond de Goncourt</u> and <u>Jules de Goncourt</u>, French novelists, published novels *Les Hommes de lettres* 'The Men of Letters', *Sœur Philomène* 'Sister Philomène', *Renée Mauperin*, *Germinie Lacerteux*, *Manette Salomon*, and great *Madame Gervaisais*, as well as founded the Académie Goncourt to foster fiction with *annual Prix Goncourt*.

1862-64: <u>Jón Árnason</u>, Icelandic collector of popular tales, wrote folk-tales and fairy-tales, which were published in the two-volume *Íslenskar þjóðsögur og ævintýri* 'Icelandic Legends'.

1862-90: <u>Alphonse Daudet</u>, French writer, produced theatrical pieces, notably *L'Arlésienne* 'A Woman from Arles'; sketches and short stories including *Lettres de mon moulin* 'Letters from my Mill' and *Tartarin de Tarascon*, continuing with *Tartarin sur les Alpes* and *Port Tarascon*; then *Le Petit chose* 'Young What's His Name'; and naturalistic novels such as *Fromont jeune et Risler aîné* 'Fromont Junior and Risler Senior', *Le Nabab* 'The Nabob', *Sapho*, and *L'Immortel* 'The Immortal One'.

1863-73: <u>Jules Verne</u>, French novelist, became worldwide known for his novels including *Cinq semaines en ballon* 'Five Weeks in a Balloon', *Voyage au centre de la terre* 'A Journey to the Centre of the Earth', *De la terre à la lune* 'From the Earth to the Moon', *Vingt mille lieues sous les mers* 'Twenty Thousand Leagues Under the Sea', and *Le Tour du monde en quatrevingts jours* 'Around the World in Eighty Days'.

1865-88: <u>Sully-Prudhomme</u>, French poet, wrote in Parnassian style pieces including *Stances et poèmes* 'Stanzas and Poems', *Les Épreuves* 'Proofs', *Croquis italiens* 'Sketches of Italy', *Impressions de la guere* 'Impressions of War', *Les Destins* 'Destinies', and *La Révolte des fleurs* 'The Flowers' Revolution'; as well as didactic poems *La Justice* 'Justice', and *Le Bonheur* 'Happiness'.

1865-92: <u>Bogdan Petriceicu Hasdeu</u>, Romanian writer and philologist, produced remarkable works such as *Historical Archive of Romania*; philological review *Trajan's Column*; 2-volume *Words from Ancestors*; *Critical History of the Romanians*; an encyclopaedic dictionary of the Romanian language entitled *Etymologicum Magnum Romaniae* that remained unfinished; dramas *Razvan and Vidra*, and *The Princess Ruxandra*; he also developed the *theory of words' circulation*; and wrote *Sic Cogito*, a theoretical work on spiritism as a philosophy.

1865-94: <u>Algeron Charles Swinburne</u>, English poet and critic, was author of drama *Atalanta in Calydon*, series *Poems and Ballads*, then *Songs before Sunrise*, Arthurian romance *Tristram of Lyonesse*, novel *Love's Cross Currents*, as well as *Essays and Studies*, and *Studies in Prose and Poetry*.

1866-80: <u>Fyodor Mikhailovich Dostoevsky</u>, Russian novelist, produced great works including *Crime and Punishment*, *The Idiot*, *The Devils*, and *Brothers Karamazov*, rejecting socialism for Russian Orthodoxy; he became widely recognized as a deep thinker.

1867-76: <u>Hristo Botev</u>, Bulgarian poet, wrote poems such as *Na proshtavene* 'At Farewell', *Haiduti* 'Hajduks', *Borba* 'Struggle', *Moyata molitva* 'My Prayer', *V mehanata* 'In the Tavern'; poetic works in the book *Song and Poems*; and the great *Obesvaneto na Vasil Levski* 'The Hanging of Vasil Levski'.

1867-1903: <u>Émile Zola</u>, French novelist, produced great novels beginning with *Thérèse Raquin*; continuing with 20-volume *Les Rougon-Macquart*, including *Nana*, *Germinal*, *La Terre* 'The Earth', and *La Bête Humaine* 'The Beast in Man'; and ending with *L'Argent* 'Money', *La Débâcle* 'The Downfall', *Le Docteur Pascal* 'Doctor Pascal', *Fécondité* 'Fruitfulness', *Travail* 'Work', and *Vérité* 'Truth'.

1869-93: <u>Paul Verlaine</u>, French poet, was author of pieces such as *Fêtes galantes* 'Gallant Parties', *La Bonne chanson* 'The Pretty Song', *Romances sans paroles* 'Romances Without Words', *Sagesse* 'Wisdom', *Parallèlement* 'In Parallel', *Amour* 'Love', *Poètes maudits* 'Accused Poets', *Louis Leclerc*, *Le Poteau* 'The Stake', *Liturgies intimes* 'Intimate Liturgies', and *Élégies*.

1870-83: <u>Mihail Eminescu</u>, Romanian greatest poet, perfected the lyric verse such in *Călin (Pages from a Fairy Tale)*; *Oh Mother*; *The Satires* 'Epistle-satires'; metaphysical *From that Star*; and especially the long poem *Luceafărul*

'The Evening Star'; as well as prose including *The Tear Drop Prince*; *Empty Genius*; *Wretched Dionis*; and *Caesara*; expressing the Latin structure and vocabulary of literary Romanian language.

1870-1900: <u>José Maria de Eça de Queiroz</u>, Portuguese writer in realist style and founder of the Naturalist School, produced novels such as *O Mistério da Estrada de Sintra* 'The Mistery of the Sintra Road', realist *O Crime do Padre Amaro* 'The Sin of Father Amaro', celebrated *O Primo Basílio* 'Cousin Bazilio', then *A Relíquia* 'The Relic', masterpiece *Os Maias* 'The Maias', and *A Ilustre Casa de Ramires* 'The Noble House of Ramires'.

1872-84: <u>Mark Twain</u>, US writer and journalist, was author of works such as *Roughing It, The Prince and the Pauper, A Connecticut Yankee in King Arthur's Court*, as well as his masterpieces *Tom Sawyer*, and *Huckleberry Finn*.

1872-96: <u>William Morris</u>, English craftsman, poet and socialist, published his works *Love is Enough, or The Freeing of Pharamond, Three Northern Love Songs*, four-volume epic *The Story of Sigurd the Volsung and the Fall of the Nibelungs*, prose romances *The Dream of John Ball*, and *News from Nowhere*, story-telling *The House of the Wolfings, The Roots of the Mountains* and *The Story of the Glittering Plain*, book of verse *Poems by the Way*, and further prose romances *The Wood beyond the World, The Well at the World's End, The Water of the Wondrous Isles* and *The Story of the Sundering Flood*.

1872-1907: <u>August Strindberg</u>, Swedish dramatist and novelist, was author of major play *Mäster Olof; Gustav Vasa*, and *Erik XIV*; novels *Röda rummet* 'The Red Room', and *Hemsöborna* 'The People of Hemsö'; psychological analyses *Fadren* 'The Father', *Fröken Julie* 'Miss Julie', and *Fordringsägare* 'The Creditors'; as well as trilogy *Till Damaskus* 'To Damascus', and play *Spöksonnaten* 'The Ghost Soonata'.

1875-99: <u>Stéphane Mallarmé</u>, French poet, was one of the leaders of *Symbolist school*, and became well-known by his poem *L'Après-midi d'un faune* 'A Faun's Afternoon', and also by works such as *Les Dieux antiques* 'The Ancient Gods', *Vers et prose*, and *Poésies*.

1875-1916: <u>Henry James</u>, US novelist, wrote works including *Roderick Hudson, The American*, critical study *French Poets and Novelists, Daisy Miller, Washington Square, Portrait of a Lady*, essay *On the Art of Fiction, Princess Casamassima, The Bostonians, The Tragic Muse, Terminations, The Spoils of Poynton, What Maisie Knew, The Two Magics, The Turn of the Screw, The Awkward Age, The Wings of a Dove, The Ambassadors, The Golden Bowl, The American Scene, The Altar of the Dead, A Small Boy and Others, Notes of a Son and a Brother*, and unfinished *The Middle Years*.

1877-1904: <u>Koizumi Yakumo</u> (Patrick Lafcadio Hearn) Japanese writer, became known by his work *Gateway to the Tropics*; continued with Louisiana works (in the USA) - *Gumbo Z'herbes: A Dictionary of Creole Proverbs, La Cuisine Créole*, novela *Chita: A Memory of Last Island*; then with writings (in Japan) - book *Glimpses of Unfamiliar Japan, Japanese Lyrics*, and legends and ghost stories such as *Kwaidan: Stories and Studies of Strange Things*.

1878-1904: <u>Thomas Hardy</u>, English novelist, poet and dramatist, wrote *The Return of the Native*; *The Mayor of Casterbridge*; *Tess of the D'Urbervilles*; *Wessex Poems*; *Winter Words*; and *The Dynasts*.

1878-1940: <u>Rabindranath Tagore</u>, Indian poet and philosopher, was author of the volume of poetry *A Poet's Tale*, novel *Karuna*, drama *The Tragedy of Rudachandra*; major works *Binodini*, poems *The Crescent Moon*, spiritual verse *Gitanjali* 'Song Offerings', known play *Chitra*, as well as *My Reminiscences*, and *My Boyhood Days*.

1879-1904: <u>Gabriele d'Annunzio</u>, Italian writer, adventurer and political leader, became known for works such as *Primo vere* 'In Early Spring'; 'Romances of the Rose', a trilogy comprising novels *Il Piacere* 'The Child of Pleasure', *L'Innocente* 'The Intruder', and *Il Trionfo della morte* 'The Triumph of Death'; tragedies *La Gioconda*, and *Francesca da Rimini*; and his great play *La figlia di Jorio* 'The Daughter of Jorio'.

1879-1910: <u>Ion Luca Caragiale</u>, Romanian playwright and short-story writer, produced the popular plays *Mr Leonida*, *A Sormy Night*, and *A Lost Letter*, the peasant drama *The False Accusation*, and prose works *An Easter Torch*, *The Sin*, and fantasy piece *Kir Ianulea*.

1879-1911: <u>Bram Stoker</u>, Irish writer, was author of works including *Personal Reminiscences of Henry Irving*; novels *The Jewel of the Seven Stars*, *The Lady of the Shroud*, and *The Lair of the White Worm*; and especially the classic vampire story *Dracula*, by which he became a celebrity.

1879-1914: <u>Anatole France</u>, French writer, produced the volume of stories *Jocaste et le chat maigre* 'Jacosta and the Thin Catt'; novel *Le Crime de Sylvestre Bonnard*, and other novels; critical studies and the like, such as the Parnassian *Le Livre de mon ami* 'My Friend's Book' as a picture of childhood happiness; satirical and sceptical works including *Les Opinions de Jérôme Coignard*; *Île des pingouins* 'Isle of Penguins'; fable *Les Dieux ont soif* 'The Gods are Thirsty'; and satire *La Révolte des anges* 'The Angels' Revolt'.

1880-92: <u>Guy de Maupassant</u>, French novelist, became known by novels including *Boule de suif* 'Ball of Tallow', *Le Horla*, *La Peur* 'The Fear', and full-length novels such as *Une Vie* 'A Woman's Life', and *Bel-Ami*.

1883-1912: <u>Emily Pauline Johnson</u>, Canadian writer, was author of poems such as *The Song My Paddle Sings*, *My Little Jean*, *A Cry from an Indian Wife*; volume of poetry *The White Wampum*; and the collection of poems *Flint and Feather*.

1883-1916: <u>Emile Verhaeren</u>, Belgian poet, playwright and founder of the school of *Symbolism*, produced the collection of poems *Les Flamandes*; then *Les Soirs*, *Les Débâcles*, and *Les Flambeaux noirs*; play *Les Aubes*; poetry books *Les Heures Claires*, *Les Heures d'Après-midi*, and *Les Heures du Soir*; tragedy *Hélène de Sparte*; and finally *Les ailes rouges de la guerre*.

1883-1919: <u>Duiliu Zamfirescu</u>, Romanian novelist and poet, wrote novels such

198

as *Facing Life*, masterpiece *Life in the Country*, *At War*, and *Betterments*; as well as poems entitled *Winter*, *Pagan Hymns*, *New Poems*, and *On the Black Sea*.

1884-1910: <u>Henryk Sienkiewicz</u>, Polish novelist, became a great literary figure due to his classical works including trilogy *Ogniem I mieczem* 'With Fire and Sword', *Potop* 'The Deluge', *Pan Wolodyjowski* 'Pan Michael'; and also *Rodzina Polanieckich* 'Children of the Soil'; and the well-known *Quo Vadis?*.

1885-1936: <u>William Butler Yeats</u>, Irish poet, published works such as *The Dublin University Review*, *Mosada: A Dramatic Poem*, *Fairy and Folk Tales of the Irish Peasantry*, *The Wanderings of Oisin and Other Poems*, *The Countess Kathleen*, *The Celtic Twilight*, *The Land of Heart's Desire*, *Poems*, 8-volume *The Collected Works in Prose and Verse*, *The Green Helmet and Other Poems*, *Responsibilities*, *The Wild Swans at Coole*, *A Vision*, *Michael Robartes and the Dancer*, *The Tower*, *The Winding Stair*, and *The Oxford Book of Modern Verse*.

1886-1912: <u>Barbu Stefanescu Delavrancea</u>, Romanian philologist and writer, developed the *Etymologicum magnum Romaniae*, started by BP Hasdeu, and published historical writings and pieces including historical trilogy *The Sunset*, *The Snowstorm*, and *Venus*, as well as the pieces *Sic Cogito*, and *Hagi-Tudose*.

1886-1916: <u>Anton Pavlovich Chekhov</u>, Russian dramatist and short-story writer, published works including *Pёstrye Rasskazy* 'Motley Stories', *Ivanov*, *Medved* 'The Bear', *Predlozheniye* 'A Marriage Proposal', *Leshy* 'The Wood Demon', *Chayka* 'The Seagull', *Dyadya Vanya* 'Uncle Vanya', *Vishnyovy Sad* 'The Cherry Orchard', and *Tri Sestry* 'The Three Sisters'.

1886-1936: <u>Rudyard Kipling</u>, English writer, created successful satirical verses *Departmental Ditties*; short stories *Plain Tales from the Hills*, and *Soldiers Three*; an attempt at a full-length novel *The Light that Failed*, collections of verse *Barrack Room Ballads*, and *The Seven Seas*; short stories *Many Inventions*, and *The Day's Work*; classic animal stories *Jungle Books*, semi-autobiographical *Kim*; children's classic *Just So Stories*; verse collection *The Five Nations*; and finally *Puck of Pook's Hill*, *Rewards and Fairies*, *Debits and Credits*, as well as autobiographical *Something of Myself*.

1887-1906: <u>Arthur Conan Doyle</u>, Scottish writer of detective stories and historical romances, produced historical romances *Micah Clarke*, *The White Company*, *Brigadier Gerard*, and *Sir Nigel*; novel *Rodney Stone*; pamphlet *The War in South Africa*; famous modern detective *A Study in Scarlet*; serial *The Adventures of Sherlock Holmes*, and books *The Sign of Four*, and *The Hound of the Baskervilles*.

1887-1908: <u>Henry Lawson</u>, Australian writer and poet, became known for his poem *A Song of the Republic*; then *The Wreck of the Derry Castle*, and *Golden Gully*; successful prose collection *While the Billy Boils*; works on the Australian bush such as the desolate *Past Carin*; sketch story *On the Edge of A Pain*; and poem *One Hundred and Three*.

1888-94: <u>Oscar Fingal Wilde</u>, Irish playwright, novelist, essayist, poet and wit, was author of classic children's stories *The Happy Prince and Other Tales*; novel *The Picture of Dorian Gray*; fairy stories *A House of Pomegranates*; *Lord Arthur Savile's Crime and Other Stories*; play *The Duchess of Padua*; dramatic-like works *Lady Windermere's Fan*, *A Woman of No Importance*, and *An Ideal Husband*; as well as his masterpiece *The Importance of Being Earnest*; and *Salomé*.

1889-1910: <u>Maurice Maeterlinck</u>, Belgian dramatist, published the volume of poetry *Les Serres chaudes* 'The Greenhouses'; prose plays *La Princesse Maleine* 'The Princess Maleine'; *Pelléas et Mélisande*; *Joyzelle*; *Marie-Magdeleine*; and popular writing *La Vie des abeilles* 'The Life of the Bee'.

1889-1912: <u>Gerhart Johann Robert Hauptmann</u>, German dramatist and novelist, produced the play *Vor Sonnenaufgang* 'Before Dawn'; then *Die Weber* 'The Weavers', introducing a theatrical phenomenon of 'collective hero'; *Florian Geyer*, marking transition to a mixture of fantasy and naturalism; *Die Versunkene Glocke* 'The Sunken Bell', and *Rose Bernd*; plays *Der Biberpeltz* 'The Beaver Coat', and *Der rote Hahn* 'The Conflagration'; as well as novels *Der Narr in Christo: Emanuel Quint* 'The Fool in Christ: Emanuel Quint', and *Atlantis*.

1889-1925: <u>Edith Somerville</u>, Irish novelist, wrote novels *An Irish Cousin*, *The Real Charlotte*, and *Some Experiences of an Irish R.M.*; then two sequels *Further Experiences...*, and *In Mr Knox's Country*; and latter *Irish Memoirs*, and *The Big House at Inver*.

1889-1934: <u>George Bernard Shaw</u>, Irish dramatist and critic, became well-known by works such as *Fabian Essays*, *The Quintessence of Ibseninsm*, *The Perfect Wagnerite*, *Widower's Houses*, *Mrs Warren's Profession*, *Arms and the Man*, *Candida*, *Three Plays for Puritans (The Devil's Disciple, Caesar and Ceopatra, and Captain Brassbound's Conversation)*, *Man and Superman*, *John Bull's Other Island*, *Major Barbara*, *The Doctor's Dilemma*, *Getting Married*, *Misalliance*, *Androcles and the Lion*, *Pygmalion*, and *Common Sense About the War*; dramas *Heartbreak House*, *Back to Methuselah*, and *Saint Joan*; prose works *The Intelligent Woman's Guide to Socialism and Capitalism*, and *The Black Girl in search of God*; as well as plays *The Apple Cart*, *Too True to Be Good*, and *The Simpleton of the Unexpected Isles*.

1890-1936: <u>Knut Hamsun</u>, Norwegian novelist, published his works including novel *Sult* 'Hunger', followed by *Mysterier* 'Mysteries', lyrical *Pan*, masterpiece *Markens grøde* 'Growth of the Soil', and unfinished *Ringen slutlet* 'The Circle is Closed'.

1891-1918: <u>Selma Ottiliaa Lovisa Lagerlöf</u>, Swedish novelist, wrote the novel *Gösta Berlings saga* 'The Story of Gösta Berling', trilogy *The Rings of the Lowenskolds*, childern's classic *Nils Holgerssons underbara resa genom Sverige* 'The Wonderful Adventures of Nils', as well as social-moral *Antikrists Mirakler* 'The Miracles of Anti-Christ', and *Bannlyst* 'The Outcast'.

1893-98: <u>George Cosbuc</u>, Romanian poet, produced *Ballads and Pastorals*; *We Demand Land*; and *Life's Struggle*; *The Mother*; *Zamfira's Wedding*; *El Zorab*; stories assembled in *Verses and Prose*; as well as collections *Threads of Spun-yarn*, and *Ballades and Idylls*.

1895-1923: <u>Maxim Gorky</u>, Russian novelist, was author of the story *Chelkash*; transitional Romanticism-Realism *Foma Gordeyev*, better-known play *Na dne* 'The Lower Depths'; as well as autobiographical trilogy *Detstvo* 'My Childhood', *Vlyudakh* 'In the World', and *Moi universitety* 'My University'.

1895-1945: <u>Herbert George Wells</u>, English novelist, short-story writer and popular historian, produced works such as *The Time Machine, The Invisible Man, The War of the Worlds, Love and Mr Lewisham, The First Men in the Moon, Kipps, The History of Mr Polly, Mr Britling Sees It Through, The Outline of History, Men Like Gods, The World of William Clissold, The Shape of Things to Come, Experiment in Autobiography,* and *Mind at the End of its Tether*.

1895-1951: <u>Paul Valéry</u>, French poet and writer, was author of poetry and poems *La Jeune parque* 'The Young Fate', and *Charmes ou poèms*; also prose works *Soirée avec M. Teste* 'An Evening with Mr Teste', and several aesthetic studies, such as *Eupalinos ou l'architecte* 'Eupalinos, or the Architect', *L'Âme et la danse* 'Dance and the Soul'; and the late short play *Le Solitaire* 'The Solitary Man'.

1897-1929: <u>John Galsworthy</u>, English novelist and playwright, produced the collection of short stories *From the Four Winds*; novels *The Island Pharisees, The Country House, Fraternity,* and *The Patrician*; plays *Strife, Justice, The Skin Game, A Bit o'Love* and *Loyalties*; also two cycles of celebrated series *Forsyte Saga*.

1898-1934: <u>George Bacovia</u>, Romanian Symbolist poet, started with poems *It's Raining, Sonnet, Winter Picture,* and *Neurosis*; then *Purple Twilight, Yellow Sparks, Pieces of Night, Decembre,* and *Spring Notes*; continued with volumes of elegant-melancholic *Lead*, then *With You*, and *Bourgeois Stanza*; and ended with last *Poems*; all of them exercising great influence upon younger contemporaries.

1899-1912: <u>Endre Ady</u>, Hungarian poet and journalist, broke away from prevailing Hungarian conservative poetry, using Symbolist procedures, and was author of the incisive collection *Uj versek* 'New Verses', and other poems such as *Vér és arany* ' Blood and Gold', *A Minden-Titkok versei* 'The Poems of All Secrets', and *A menekülő Élet* 'The Fleeing Life'.

1899-1919: <u>Władysław Stanisław Reymont</u>, Polish novelist, was author of works including *Ziemia obiecana* 'The Promised Land' dealing with urban life in industrial town Łódź, followed by his masterpiece tetralogy *Chłopi* 'The Peasants' studying rural life, then *Komediantka* 'The Comédienne' and historical trilogy *Rok 1794* 'The Year 1794'.

1900-23: <u>Rainer Maria Rilke</u>, Austrian lyric poet, wrote *Vom lieben Gott und*

201

Anderes 'Stories of God', *Das Stundenbuch* 'Poems for the Book of Hours', *Auguste Rodin, Neue Gedichte* 'New Poems', *Die Aufzeichnungen des Malte Laurids Brigge* 'Journal of My Other Self'; and his major works *Die Sonnette an Orpheus* 'Sonnets to Orpheus', and *Duineser Elegien* 'Duino Elegies'.

1902-50: Andŕe Paul Guillaume Gide, French novelist, writer and diarist, produced novels *L'Immoraliste* 'The Immoralist', *La Porte étroite* 'Strait is the Gate', *Les Caves du Vatican* 'The Vatican Swindle', *La Symphonie pastorale* 'Two Symphonies', and *Les Faux-monnayeurs* 'The Counterfeiters'; founded the magazine *La Nouvelle Revue Française*; and also wrote *Journals* as an essential supplement to autobiography *Si le grain ne meurt* 'If It Die...'.

1902-52: Mihail Sadoveanu, Romanian novelist and story writer, was author of novel *The Potcoava Brothers*, books *Soimii* 'The Falcons', *Suppressed Pains, Stories from the War*, and *Roving Times*; followed by works on Romania's medieval and early modern history such as *The Soimaresti Family, The Jderi Brothers*, and *Under the Sign of the Crab*; as well as on contemporary history including *A Mill Was Floating down the Siret*, and *The Hatchet*; then *Thomas Sunday*, cycle *His Lordship's Men*, and political novel *Mitrea Cocor*; as well as *The Flowers' Lure*, and novel *Nicoara Potcoava*.

1903-13: Jack London, US writer, produced popular novels, such as *The Call of the Wild, The Sea-Wolf,* and *White Fang; The Iron Heel, Martin Eden*, and *John Barleycorn*.

1903-38: Allama Muhammad Iqbal, Pakistani philosopher and poet, became known by his poetry book *Asrar-e-Khudi* 'Secrets of the Self'; and other books of poetry including *Rumuz-i-Bekhudi* 'Hints of Selflessness', *Payam-i-Mashriq* 'The Message of the East', and *Zabur-i-Ajam* ' Persian Psalms'.

1904-26: Isobel Marion Dorothea Mackellar, Australian poet and fiction writer, was author of the poem *My Country*; and volumes of collected verse *The Closed Door; The Witch Maid, and Other Verses; Dreamharbour*; and *Fancy Dress*.

1904-30: Luigi Pirandello, Italian dramatist, novelist and short-story writer, produced powerful and realistic novels and short stories, such as *Il Fu Mattia Pascal* 'The Late Mattia Pascal', *Si Gira* 'Shoot!', and plays *Sei personaggi in cerca d'autore* 'Six Characters in Search of an Author', *Enrico IV* 'Henry IV', and *Come Tu Mi Vuoi* 'As You Desire Me'.

1904-55: Hermann Hesse, Swiss novelist and poet, published novel *Peter Camenzind*; prose *Rosshalde; Knulp; Demian; Narziss und Goldmund* 'Death and the Lover'; *Steppenwolf; Das Glasperlenspiel* 'The Glass Bead Game'; and poetry *Die Gedichte* 'Hours in the Garden and Other Poems', as well as *Beschwörungen* 'Affirmations'.

1905-37: Gheorghe Toparceanu, Romanian poet, short story writer and humorist, was author of volumes of poetry *Ballads, Merry and Sad, Original Parodies*, and *Bitter Almonds*; celebrated poems *The Ballad of a Tiny Cricket, Fall Rhapsodies, The Bullet Train*, and *The Crow*; and prose *Memoires from the Battle of Turtucaia, Letters with No Address*, and *Pirin-Planina*.

1907-22: <u>James Augustine Aloysius Joyce</u>, Irish writer and poet, produced books *Chamber Music*, and *Dubliners* as a collection of short stories, and moreover the seminal novel *Ulysses*.

1907-43: <u>William Henry Davies</u>, Welsh poet, published *A Soul's Destroyer*; *The Autobiography of a Super-tramp*; autobiographical *Beggars*; *The True Traveller*; *A Poet's Pilgrimage*; *Later Days*; also *Adventures of Johnny Walker*, and *Collected Poems*.

1907-50: <u>Tudor Arghezi</u>, Romanian writer, became known for his works including *Ode to Mankind*; masterpiece poems *Suitable Words, Wooden Icons, Flowers of Mildew, The Black Gate*; and novels *Tablets from the Land of Kuty, Swiftian Stories*, and *Our Lord's Mother's Eyes*.

1908-46: <u>Jules Romains</u>, French writer, was author of poems *La Vie unanime* 'The Unanimous Life', *Manuel de deification* 'A Treatise on Deification', novels *Mort de quelqu'un* 'The Death of a Nobody' and *Les Copains* 'The Friends', successful play *Knock, ou le triomphe de la médicine* 'Doctor Knock', poetry including *Chants des dix années 1914-1924* 'Songs of the Ten Years 1914-24' and *L'Homme blanc* 'The White Man', as well as his great 27-volume cycle *Les Hommes de bonne volonté* 'Men of Good Will'.

1909-20: <u>Emilio Filippo Tommaso Marinetti</u>, Italian writer, published, in the newspaper *Le Figaro*, his *Manifesti del Futurismo* 'Futurist Manifesto', glorifying the war, machine age-speed, and a sort of 'dynamism' as a revolt against tradition; then he wrote *Teatro sintetico futuristo* 'The Synthetic Futurist Theatre', and 4-volume *Manifesti del Futurismo* 'Manifesto of Futurism'.

1909-31: <u>Ion Minulescu</u>, Romanian avant-garde poet, novelist, literary critic and playwright, produced poems such as *Songs for Later On*, and *Verses for Everyone*; prose *The House with Orange Windows*, and novel *Flunking in Romanian Language*; as well as plays including *The Storks Are Leaving, Lulu Lupescu, The Sentimental Mannequin*, and *The Wingless Dove*.

1909-41: <u>Roger Martin du Gard</u>, French novelist, wrote *Devenir* 'Becoming', founded the *Nouvelle revue française*, then published novels including *Jean Barois, Vieille France* 'The Postman', and especially the eight-novel series *Les Thibault* 'The Thibaults' dealing with family life during first decades of 20th century.

1909-46: <u>Guillaume Apollinaire</u>, French poet, produced works such as *L'Enchanteur pourrissant* 'The Decaying Magician', *Le Bestiaire, Alcools*, and *Calligrammes*; play *Les Mamelles de Tirésias* 'The Breasts of Tiresias' coining term 'surrealist'; and Modernist manifesto *L'Esprit nouveau et les poètes* 'The New Spirit and the Poets'.

1909-62: <u>William Carlos Williams</u>, US poet, novelist and cultural historian, was author of volumes including *Poems, The Tempers*, and *Sour grapes*; followed by *Spring and All, The Great American Novel, In the American Grain*, and *White Mule*; trilogy *In the Money, The Build-Up*, and masterpiece

Paterson; as well as *The Collected Later Poems, The Collected Earlier Poems*, and *Pictures from Breughel*.

1910-21: Ramón López Velarde, Mexican poet, wrote *La sangre devota* 'The Pious Blood', major book *Zozobra* 'Sinking'; also essays *Novedad de la Patria*, and unfinished book *El son del corazón* 'The sound of the heart'.

1910-33: Romain Rolland, French musicologist and writer, published works including *Beethoven*, ten-volume novel cycle *Jean-Christophe, Au dessus de la mêlée* 'Above the Fray', and another novel cycle *L'Âme enchantée* 'The Enchanted Spirit'.

1910-46: Ivan Alekseyevich Bunin, Russian novelist and story writer, became well-known by his short novels *The Village*, and *Dry Valley*; then *Coursed Days*; autobiographical novel *The Life of Arseniev*; and the book of short stories *Dark Avenues*.

1911-36: Mehmet Akif Ersoy, Turkish poet and author, wrote the collection of 44 poems *Safahat*, the national anthem *İstiklâl Marşi* 'The March of Independence', and also *Hakkın Sesleri* 'Voices of God', *Gölgeler* 'Shadow', and *Kur'an'dan Ayet ve Hadisler* 'Ayat and Hadith from the Koran'.

1912-38: Liviu Rebreanu, Romanian novelist, playwright, short story writer and journalist, was author of short stories *The Hooligans, Confession*, and *Resentfulness*; modern novels *Ion* 'John', *The Little King, The Revolt*, and *The Gorilla*; psychological novels *Forest of the Hanged, Adam and Eve*, and *Ciuleandra*; and plays *The Quadrille, The Envelope*, and *The Apostles*.

1912-41: Virginia Woolf, English novelist, critic and essayist, wrote novels *The Voyage Out, Night and Day, Mrs Dalloway, To the Lighthouse, Orlando, The Waves, The Years, Three Guineas*, and *Between the Acts*; as well as the outstanding *Jacob's Room*.

1913-22: Marcel Proust, French novelist, was author of the 13-volume major work *À la recherche du temps perdu* 'Remembrance of Time Past', including a series of autobiographical novels such as *Du côté de chez Swann* 'Swann's Way', *À l'ombre des jeunes filles en fleur* 'Within a Budding Grove', *Le Côté de Guermantes* 'The Guermantes' Way', and *Sodome et Gomorrhe* 'The Cities of the Plain', as well as other two-volume works published posthumously.

1913-24: Franz Kafka, Austrian novelist, published influential novels *Prozess* 'The Trial', *Das Schloss* 'The Castle', and *Amerika*, portraying society as a pointless, schizophrenically rational organization where the individual has strayed.

1913-62: Robert Lee Frost, US lyric poet, produced works such as *A Boy's Will*, and *North of Boston*; then volumes of poetry *West-Running Brook, A Witness Tree, Steeple Bush*, and *In the Clearing*.

1916-20: Hugo Ball, German artist, Hans Arp, Alsatian sculptor and poet, and Tristan Tzara, Romanian poet, promoted a literary-artistic movement that was opposing war and cultural values, but espousing anarchic individualism and freedom from artistic convention.

1917-44: <u>Max Jacob</u>, French Cubist writer, mystic, astrologer, artist and monk, produced the collection of prose-poems *Le Cornet de dés* 'The Dice Cup', and mystical works, notably *L'Homme de cristal* 'The Crystal Man'.

1918-25: <u>Lu Xun</u>, Chinese writer, was author of short stories such as *Diary of a Madman*, successful *The True Story of Ah Q*, and cycles *Cry*, and *Hesitation*.

1918-30: <u>Vladimir Vladimirovich Mayakovsky</u>, Russian poet and playwright, produced the play *Misteriya-Buff* 'Mystery-Bouffe', long poem *150,000,000*, *Pro eto* 'About This', satirical plays including *Klop* 'The Bedbug', and *Banya* 'The Bath-House', *Vladimir Ilych Lenin, Khorosho!* 'Good!', and unfinished *Vo ves golos*.

1918-38: <u>César Abraham Vallejo Mendoza</u>, Peruvian poet, writer and playwright, published his books of poetry including *Los Heraldos Negros* 'The Black Messengers', *Trilce*, *Fabla Salvaje* 'Wild Language', novel *El tungsteno*, as well as *España, Aparta de Mí Este Cáliz* 'Take This Chalice from Me', and *Poemas Humanos* 'Human Poems'.

1918-45: <u>Bertolt Friedrich Brecht</u>, German playwright and poet, became known by his works including *Trommeln in der Nacht* 'Drums in the Night', *Mann ist Mann* 'A Man's a Man', *Dreigroschenoper* 'The Three-penny Opera', *Mutter Courage und ihre Kinder* 'Mother Courage and her Children', and *Furcht und Elend des dritten Reiches* 'Fear and Loathing under the Third Reich'.

1918-55: <u>Thomas Mann</u>, German novelist, wrote notable works such as *Betrachtungen eines Unpolitischen* 'Reflections of a Non-political Man', *Der Zauberberg* 'The Magic Mountain', *Achtung Europa! Deutsche Hörer!* 'Listen Germany: Twenty-Five Messages to the German People over the BBC'; great work *Doktor Faustus*; and comic novel *Bekenntnisse des Hochstaplers Felix Krull* 'Confessions of Felix Krull, Confidence Man'.

1918-59: <u>André Maurois</u>, French novelist and biographer, was author of works including *Les Silences du Colonel Bramble* 'The Silences of Colonel Bramble', *Les Discours du Docteur O'Grady* 'The Speeches of Dr O'Grady'; biographies of *Ariel, Disraeli, Voltaire, À la recherche de Marcel Proust* 'In Search of Marcel Proust', and *La vie de Sir Alexander Fleming* 'The Life of Sir Alexander Fleming'.

1918-70: <u>Ivan (Ivo) Andrić</u>, Yugoslav novelist and short story writer, published his literary such as *Ex Ponto, Unrest, The Pasha's Concubine and other Tales, Bosnian Chronicle, The Bridge on the Drina, The Woman from Sarajevo, The Vizier's Elephant*, and also wrote *The Damned Yard*.

1920-57: <u>Rómulo Ángel del Monte Carmelo Gallegos</u>, Venezuelan novelist and politician, produced works including *El último Solar* 'Reinaldo Solar', and novels *Doña Bárbara, Cantaclaro, Canaima, Pobre negro, La rebelión*, and *El último patriota*.

1920-58: <u>Thomas Stearns Eliot</u>, US-born British poet, critic and dramatist, wrote works of critics, e.g. *The Sacred Wood, Homage to Dryden, The Use of Poetry and the Use of Criticism, Elizabethan Essays*, and *On Poetry and*

Poets; social comments, e.g. *After Strange Gods, Essays Ancient and Modern, The Idea of a Christian Society*, and *Notes towards the Definition of Culture*; volume of essays *For Lancelot Andrewes*; religious plays *The Rock*, and *Murder in the Cathedral*; and dramas *The Cocktail Party, The Confidential Clerk*, and *The Elder Statesman*.

1921-62: Aldous Leonard Huxley, English novelist and essayist, produced novels *Crome Yellow, Antic Hay, Those Barren Leaves*, and *Point Counter Point*; essays including *Proper Studies*; his famous novel *Brave New World*; *Eyeless in Gaza*, and *After Many a Summer*, pointing the way to *Time must have a Stop*; study in sexual hysteria *The Devils of Loudun*; *The Doors of Perception*, and *Heaven and Hell*, exploring a controversial short cut to mysticism; and the optimistic Utopian novel *Island*.

1921-66: Ilya Grigorevich Ehrenburg, Russian novelist and journalist, was author of *The Extraordinary Adventures of Julio Jurenito, The Fall of Paris, The Storm, The Thaw*, and finally *People, Years, Life*.

1921-77: Jaroslav Seifert, Czech poet and journalist, produced collections *Město v slzáck* 'City of Tears', and *Samá láska* 'All Love'; also *A Wreath of Sonnets*; as well as patriotic works *Přílba Llíny* 'A Helmet of Earth', and *Morový sloup* 'The Prague Column'.

1922-38: François Mauriac, French novelist, wrote major works including *Le Baiser au lépreux* 'The Kiss to the Leper', *Génitrix, Thérèse Desqueyroux, Le nœud de vipères* 'Vipers' Tangle', and play *Asmodée*.

1922-54: Yasunari Kawabata, Japanese writer, published short stories *Tales to hold in the Palm of your Hand*; and novels *Izu no odoriko* 'The Izu Dancer', *Yukiguni* 'Snow Country', *Sembazuru* 'Thousand Cranes', and *Yama no oto* 'The Sound of the Mountain'.

1923-30: Panait Istrati, Romanian writer of French language, was known for his works written in Romanian *The Thistles of the Bărăgan*, and *Kira Kyralina*; as well as in French *Vers l'autre flamme, confession pour vaincus*, directed against USSR's Stalinist regime; he became reputed as the 'Gorki of the Balkans'.

1923-52: Ernest Millar Hemingway, US writer of novels and short stories, produced *Three Stories and Ten Poems, Our Time, The Sun Also Rises, Men Without Women, A Farewell to Arms, Death in the Afternoon, Green Hills in Africa, For Whom the Bell Tolls, Across the River and Into the Trees*, and *The Old Man and the Sea*.

1923-54: Lion Feuchtwanger, German writer, was author of *Die hässliche Herzogin* 'The Ugly Duchess', *Jud Süss* 'Jew Süss', *Erfolg* 'Success', and part-biographies of F de Goya, and JJ Rousseau.

1924-42: Stefan Zweig, Austrian-born British writer, became known for his short stories such as *Kaleidoskop* 'Kaleidoscope', and especially novels *Der Zwang* 'Passion and Pain', and *Ungeduld des Herzens* 'Beware of Pity', all notable for their deep psychological insights.

1924-66: Pablo Neruda, Chilean poet, published *Veinte poemas de amor y una canción desesperada* 'Twenty Love Poems and a Song of Despair', *Residencia en la tierra* 'Residence on Earth', *Alturas de Macchu Picchu* 'The Heights of Macchu Picchu', *Odas elementales* 'Elementary Odes', and comprehensive *Canto General* 'Poems from Canto General'.

1925-34: <u>Kim Sowol</u>, Korean poet, wrote remarkable poem *The Azaleas*; and other poems such as *Mother and Sister, Invocation*, and *The Way*.

1926-59: <u>Eugene Gladstone O'Neill</u>, US playwright, was author of *The Great God Brown, Marco Millions, Strange Interlude, Lazarus Laughed*; the trilogy *Mourning Becomes Electra, Ah, Wilderness, Days Without End*; *The Iceman Cometh, A Moon for the Misbegotten, Long Day's Journey into Night*, as well as *A Touch of the Poet*, and *Hughie*.

1927-31: <u>Ahmed Shawqi</u>, Egyptian poet and dramatist, wrote tragedies such as *Majnun Laila* 'The Mad about Layla', *The Death of Cleopatra, 'Antara, Ali bek el-Kabeer*; poetry *The States of Arabs and the Great Men of Islam*, prose *The Markets of Gold*; and *Qambeez* 'Cambyses II'.

1927-48: <u>Mahatma Gandhi</u> (Great Soul), Indian leader and writer, produced autobiographical *The Story of My Experiment with Truth*, and works comprised in the 90-volume *The Collected Works of Mahatma Gandhi*, with a great influence for peace as a message not only for India but for entire world.

1927-57: <u>Boris Leonidovich Pasternak</u>, Russian lyric poet, novelist and translator, wrote political poems *Devyat'sot pyaty god* 'The Year 1905', on Bolshevik uprising, and autobiographical *Vtoroye rozhdeniye* 'Second Birth'; short stories *Detstvo Lyuvers* 'The Childhood of Luvers', and *Provest'* 'The Last Summer'; and famous novel *Doktor Zhivago* 'Doctor Zhivago'.

1928-37: <u>Kulap Saipradit</u>, Thai writer, was author of novels including *Luk Phu Chai* 'A Real Man', and *Songkram Chiwit* 'The War of Life; as well as the romantic masterpiece *Khang Lang Phap* 'Behind the Painting'.

1928-51: <u>André Malraux</u>, French writer, made dramatic meditation on human destiny, such as displayed in novels *Les Conquérants* 'The Conquerors', *La Condition humaine* 'Man's Fate', and *L'Espoir* 'Man's Hope'; he also wrote *La Psychologie de l'art* 'The Psychology of Art', and four-volume *Les Voix du silence* 'The Voices of Silence'.

1928-69: <u>Mikhail Aleksandrovich Sholokhov</u>, Russian novelist, was author of the masterpiece *Tikhy Don* 'And Quiet Flows the Don and The Don Flows Home to the Sea', and less valuable *Podnyataya tselina* 'Virgin Soil Up-turned and Harvest on the Don'.

1929-59: <u>William Faulkner</u>, US novelist, produced *The Sound and the Fury*; *Sartoris*; *As I Lay Dying*; *Sanctuary*; *Light in August*; *Absalom, Absalom!*; *Hamlet*; *Intruder in the Dust*; *A Fable*; *The Town*, and *The Mansion*.

1930-76: <u>Richard Ghormley Eberhart</u>, US poet, wrote almost 30 collections, including *A Bravery of Earth, Reading the Spirit, The Quarry, Fields of Grace, Selected Poems 1930-1965*, and *Collected Poems 1930-1976*.

1930-40: <u>Camil Petrescu</u>, Romanian novelist and poet, was author of remarkable novels such as *The Last Night of Love, the First Night of War*, and

Procrustes's Bed; as well as philosophical works including *The Doctrine of Substance*.

1931-40: <u>Ba Jin</u>, Chinese writer, produced a major trilogy, comprising *Jia* 'The Family', *Chun* 'Spring', and *Qiu* 'Autumn', in which he attacked the traditional family system, and became immensely popular with younger generation.

1931-60: <u>Georges Joseph Christian Simenon</u>, French novelist, created famous novels featuring the detective Jules Maigret, such as *M. Gallet décède* 'The Death of Monsieur Gallet', *Le Pendu de Saint-Pholien* 'The Crime of Inspector Maigret', and *Les Mémoires de Maigret* 'Maigret's Memoirs'.

1931-64: <u>Eduardo De Filippo</u>, Italian actor, playwright and poet, was author of works including *Ogni anno punto e da capo* 'Every Year Back from the Start'; *Natale in casa Cupiello* 'Christmas at the Cupiello's'; the well-known *Napoli Milionaria* 'The Millions of Naples'; as well as *Filumena Marturano*; and *L'arte della commedia* 'The Art of Comedy'.

1931-65: <u>George Seferis</u>, Greek poet and diplomat, published a collection of poetry entitled *Strophe* 'Turning Point' with an immediate success, *Mythistorima* 'Myth History' containing some of first free-verse Greek poems, and also collections including the three-volume *Hemerologhia Katastromatos* 'Logbook'.

1931-68: <u>Halldór Kiljan Laxness</u>, Icelandic novelist, wrote epic novels like *Salka Valka*; *Sjálfstætt fólk* 'Independent People'; *Heimsljós* 'World Light'; *Íslandsklukkan* 'Iceland's Bell'; *Atómstöðin* 'The Atom Station'; *Gerpla* 'The Happy Warriors'; *Brekkukotsannáll* 'The Fish Can Sing'; *Paradisarheimt* 'Paradise Reclaimed'; and *Kristnihald undir Jökli* 'Christianity at Glacier'.

1931-76: <u>John Betjeman</u>, English poet and writer on architecture, produced the collection of verse *Mount Zion; or, In Touch with the Infinite*; book *Ghastly Good Taste*; other collections including *Continual Dew: A Little Book of Bourgeois Verse*, *Old Lights for New Chancels*, *New Bats in Old Belfries*, *A Few Late Chrysanthemums*, and *Collected Poems*; *Cornwall*, as well as volumes *A Nip in the Air*, and *High and Low*.

1932-41: <u>Tengku Amir Hamzah</u>, Indonesian poet, was author of 50 poems, 18 pieces of lyrical prose, and collections of poems, including *Nyanyi Sunyi*, and *Buah Rindu*.

1932-47: <u>Federico García Lorca</u>, Spanish poet and playwright, wrote plays *Bodas de Sangre* 'Blood Wedding', *Yerma*, and *La Casa de Bernarda Alba* 'The House of Bernarda Alba'; gypsy songs *Canciones*, and *Romancero Gitano* 'Gypsy Ballads'; and elegiac poems *Llanto por la muerte de Ignacio Sánchez Mejías* 'Lament for the Death of a Bullfighter and Other Poems'.

1933-47: <u>Mihail Sebastian</u>, Romanian novelist and playwright, was author of novels including *Women, It's Been Two Thousand Years, The Acacia Tree City*, and *The Accident*; as well as theatrical pieces *Holiday Games, The Star without a Name, Breaking News*, and *The Island*.

1935-61: <u>John Ernest Steinbeck</u>, US novelist, started with reputed novel

Tortilla Flat, continued with works *In Dubious Battle, Of Mice and Men, The Moon is Down, The Pearl, Burning Bright, East of Eden, Winter of our Discontent*, and ended with humorous *Cannery Row*, and *The Short Reign of Pippin IV*, displaying liberal humanism and generous solidarities.

1935-92: <u>Elias Canetti</u>, Bulgarian-born Swiss and British modernist novelist, playwright, non-fiction writer and memoirist, became known for his novel *Auto - da - Fé*, study *Crowds and Power*, travelogue *The Voices of Marrakesh*, as well as memoirs *The Torch in My Ear*, also *The Play of the Eyes*, and book *The Agony of Flies*.

1938-57: <u>Jean-Paul Sartre</u>, French philosopher, dramatist and novelist, wrote the autobiographical novel *La Nausée* 'Nausea', collection of short stories *Le Mur*; plays *Les Mouches* 'The Flies', and *Huis clos* 'In Camera'; as well as *Les Mains sales* 'Crime Passional', *L'Existentialisme est un humanisme* 'Existentialism and Humanism', *L'Être et le néant* 'Being and Nothingness', and trilogy *Les Chemins de la liberté* 'The Paths of Freedom'.

1938-70: <u>Samuel Barclay Beckett</u>, Irish writer and playwright, published novels *Murphy*, and *Watt* in English; trilogy *Molloy, Malone Meurt*, and *L'Innommable* in French; plays *En attendant Godot*, and *Fin de partie* in French; then *Happy Days, Not I*, and *Ill Seen Ill Said*; as well as later short piece *Breath*.

1942-53: <u>Albert Camus</u>, French writer, produced the existentialist novel *L'Étranger*; then the work *Le Myth of Sisyphus*, on suicide; masterpiece *La Peste*; ironic *La Chute*; plays *Le Malentendu*, and *Caligula*; also political *Actuelles*.

1942-65: <u>Salvatore Quasimodo</u>, Italian poet, was author of works including *Ed è Subito Sera* 'And Suddenly It is Evening', *La Vita non è sogno* 'Life is Not a Dream', *La Terra impareggiabile* 'The Matchless Earth', and the collection *Selected Poems*.

1942-91: <u>Camilo José Cela</u>, Spanish novelist, short story writer and essayist, became known for his novels *The Family of Pascual Duarte*, and *The Hive*; travel book *Viaje a la Alcarria*; and other books including *Cristo versus Arizona*, and *San Camilo 1936*.

1944-49: <u>Jean Genet</u>, French author, wrote novels *Notre-Dame des fleurs* 'Our Lady of the Flowers', *Miracle de la rose* 'Miracle of the Rose', and *Pompes funèbres* 'Funeral Rites'; plays *Les Bonnes* 'The Maids', *Les Nègres* 'The Blacks', and *Les Paravents* 'The Screens'; poems *Les Condamnés á mort* 'Those Condemned to Death', and *Chants Secrets* 'Secret Songs'; as well as his autobiography *Le Journal du voleur* 'Thief's Journal'.

1947-83: <u>Naguib Mahfouz</u>, Egyptian novelist, became known for his books including *Midaq Alley, Palace Walk, Palace of Desire, Children of Gebelawi*, and *The Journey of Ibn Fattouma*.

1949-75: <u>Heinrich Böll</u>, German writer, published *Der Zug war pünktlich* 'The Train was on Time'; trilogy *Und sagte kein einziges Wort* 'Acquainted with the Night', *Haus ohne Hüter* 'The Unguarded House', and *Das Brot der*

209

frühen Jahre 'The Bread of our Early Years', depicting life in Germany during and after Nazi regime; then his later novels including *Gruppenbild mit Dame* 'Group Portrait with Lady', and *Die vorlorene Ehre der Katherina Blum* 'The Lost Honour of Katherina Blum'.

1949-95: <u>Nadine Gordimer</u>, South African novelist, wrote the collections *Face to Face*, and *The Soft Voice of the Serpent*; the novel *The Lying Days*; the books *Occasion for Loving*; *The Late Bourgeois World*; *A Guest of Honour*; *The Conservationist*; *Burger's Daughter*; *July's People*; and *A Sport of Nature*; as well as *None to Accompany Me*; and *Writing and Being*.

1950-80: <u>Eugen Ionesco</u>, Romanian-born French playwright, introduced a new style of drama, called the *Theatre of the Absurd*, comprising pieces such as *La Cantatrice chauve* 'The Bold Prima Donna'; *Les Chaises* 'The Chairs'; *Amédée*; *Le Tableau* 'The Picture'; *Rhinocéros*; *Jeux de massacre* 'Wipe-Out Game'; *Macbett*; *Voyages chez les morts ou Thème et variations* 'Journey Among the Dead'; and also the novel *Le Solitaire* 'The Hermit'.

1950-2003: <u>Doris May Lessing</u>, British novelist, poet, playwright and short story writer, became known for her published works including *The Grass Is Singing, The Golden Notebook, Shikasta, Canopus in Argos: Archives, The Good Terrorist, The Fifth Child*, and *The Grandmothers: Four Short Novels*.

1951-88: <u>Dame Muriel Sarah Spark</u>, Scottish novelist, short-story writer and poet, published short story *The Seraph and the Zambesi*; then *The Comforters*; *Memento Mori*; *The Ballad of Peckham Rye*; and *The Bachelors*; the novel *The Prime of Miss Jean Brodie*; and *The Girls of Slender Means*; *The Mandelbaum Gate*; *The Abbess of Crewe*; *Loitering with Intent*; *The Only Problem*; and *A Far Cry from Kensington*.

1951-90: <u>Kukrit Pramoj</u>, Thai politician, scholar and writer, became known for his novels such as *Si Phaendin* 'Four Reigns', and *Lai Chiwit* 'Many Lives'; play *Rashomon*; and short stories such as *Phuean Non*, and *Sapphehera Khadi*.

1952-61: <u>Frantz Omar Fanon</u>, Martinique-born French psychiatrist, philosopher and writer, was author of works including *Black Skin, White Masks*; *Year Five of the Algerian Revolution*; *A Dying Colonialism*; *The Wretched of the Earth*; and *Toward the African Revolution*; which became source of inspiration and manifesto for liberation struggles throughout the Third World.

1953-80: <u>Gely Abdel Rahman</u>, Sudanese poet, produced *Gsa'aid meen El-sudan* 'Poems from Sudan', *The Foreign Aid and its Influence On Sudan Independence*, poetry *Cavalery and Broken Sword, Gates of Yellow Cities*, and *Fire and Dreams of Nightingales*.

1953-2002: <u>Czesław Miłosz</u>, Lithuanian-Polish poet and prose writer, 'who with uncompromising clear-sightedness voices man's exposed condition in a world of severe conflicts', became known for his both prose, such as *The Captive Mind*; and *Modern Legends, War Essays*; and poetry including *The Unencompassed Earth*; and *The Second Space*.

1954-83: <u>Françoise Sagan</u>, French novelist, wrote pieces including *Bonjour tristesse*, *Un Certain Sourire* 'A Certain Smille', *Dans un mois, dans un an* 'Those Without Shadows', *Aimez-vous Brahms* 'Goodbye Again', *La Chamade, Le Rendez-vous manqué* 'The Missed Rendezvous'; plays *Château en Suède* 'Castle in Sweden', and *Un Piano dans l'herbe* 'A Piano on the Grass'; and finally novels *La Femme fardée* 'The Painted Lady', and *Un Orange immobile* 'The Still Storm'.

1954-90: <u>Roald Dahl</u>, British children's author, short-story writer, playwright and versifier, produced collections such as *Someone Like You, Kiss Kiss*, and *Switch Bitch*; novel *My Uncle Oswald*; children's stories *Charlie and the Chocolate Factory, Charlie and the Great Glass Elevator, James and the Giant Peach, Fantastic Mr Fox, The Enormous Crocodile, The BFG, Matilda*, and *Esio Trot*; as well as screen-plays *You Only Live Twice*, and *Chitty Chitty Bang Bang*.

1955-67: <u>Marin Preda</u>, Romanian novelist and story writer, was author of the masterpiece *Moromeţii* 'The Moromites', and also the novel *Cel mai iubit dintre pământeni* 'The Most Loved from Countrymen'.

1955-90: <u>Catherine Ann Cookson</u>, English popular novelist, inspired mainly by her deprived youth in South Tyneside, North-east England, she was a prolific writer with more than 70 books published, notably tragedy and romance, including *Mallen trilogy*, and *Tilly Trotter series*; she became well-known for her filmed *The Fifteen Streets*; the books *A Grand Man*, as a basis of the film 'Jacqueline', and *Rooney*, also adapted for screen; then *Katie Mulholland*, adapted into a stage musical; as well as the filmed *The Black Valvet Gown*.

1956-90: <u>Aleksandr Isayevich Solzhenitsyn</u>, Russian writer, produced the novel *Odin den'Ivana Denisovicha* 'One Day in the Life Ivan Denisovich', followed by other novels entitled *Rakovy Korpus* 'The Cancer Ward', and *V Kruge pervom* 'The First Circle', factual account of Stalinist terror *Arkhipelag Gulag* 'The Gulag Archipelago', also *Bodalsya telyonok s dubom* 'The Oak and the Calf', and *Kak Nam Obustroit' Rossiyu?*, translated as 'Rebuilding Russia'.

1956-94: <u>Ted Hughes</u>, English poet, wrote *The Hawk in the Rain*; *Wodwo*; *Crow*; *Cave Birds*; *Season Songs*; *Gaudete*; *Moortown*; *Remains of Elmet*; *River*; *Wolf Watching*; *Tales from Ovid*; *The Iron Man*; and *The Iron Woman*; as well as books for children, collection *Rain-Charm for the Duchy and other Laureate Poems*, and essays such as *Winter Pollen: Occasional Prose*.

1956-97: <u>Octavio Paz Lozano</u>, Mexican poet and writer, became known for his essays *The Bow and the Lyre*, and *The Labyrinth of Solitude*; poetry *The Collected Poems*; and also books *The Double Flame*, and *In Light of India*.

1957-93: <u>Dame Iris Murdoch</u>, Irish novelist, playwright and philosopher, was author of *The Sandcastle*; *The Bell*; *A Several Head*; *The Sea*; *Nuns and Soldiers*; *The Good Apprentice*; *The book and the Brotherhood*; *The Message to the Planet*; and *The Green Knight*.

1957-2007: <u>Harold Pinter</u>, English dramatist, produced *The Birthday Party*, and filmed *The Caretaker*; which were followed by television plays *The Lover*, and *The Collections*; radio and stage *The Dwarfs*, and *The Homecoming*; film-scripts *The Servant*, and *The Pumpkin*, and plays *No Man's Land*, *Betrayal*, *Party Time*, and *Moon light*; short pieces *Other Voices* shown at the National Theatre in London; political themes *One for the Road*, *Mountain Language*, and *A New World Order*; film-scripts *The French Lieutenant's Woman*, *The Handmaid's Tale*, and *The Comfort of Strangers*; and other works such as *The Go-Between*, *The Trial*, and *Sleuth*.

1958-85: <u>Oë Kenzaburo</u>, Japanese novelist, published *Shisha no Ogori* 'The Arrogance of the Dead'; *Nip the Buds, Shoot the Kids*; *Kojinteki na taiken*; *A Personal Matter*; and *Man'en gannen no futtuboru* 'The Silent Cry', as well as short novels collected as *Teach Us to Outgrow Our Madness*; and *The Crazy Iris and Other Stories of the Atomic Aftermath*.

1958-87: <u>Chinua Achebe</u>, Nigerian novelist, poet and critic, became known by his works *Things Fall Apart*; as well as novels *No Longer at Ease*, *Arrow of God*, *A Man of the People*, and *Anthills of the Savannah*.

1958-2002: <u>Claude Simon</u>, French novelist, wrote a series of books including *The Grass*; *La route des Flandres*, *Triptych*; *The Acacia: A Novel*; and *The Trolley*.

1958-2009: <u>Dario Fo</u>, Italian playwright, comedian and composer, became known for his works such as *Non tutti i ladri vengono a nuocere* 'The Virtuous Burglar'; *Gli arcangeli non giocano al flipper* 'Archangels Don't Play Pinball '; *Mistero Buffo*; *Morte accidentale di un anarchico* 'Accidental Death of an Anarchist'; *Non Si Paga! Non Si Paga!* 'Can't Pay? Won't Pay!'; as well as *Clacson, trombette e pernacchi* 'Trumpets and Raspberries'; *Il Papa e la strega* 'The Pope and the Witch'; and *Francis the Holy Jester*.

1959-80: <u>Abe Kobo</u>, Japanese novelist and playwright, produced novels *Daiyon Kampyoki* 'Inter Ice Age Four', *Suna no onna* 'The Woman in the Dunes', and *Mikkai* 'Secret Rendezvous'.

1959-95: <u>Günter Wilhelm Grass</u>, German novelist, wrote the novel *Die Blechtrommel* 'The Tin Drum'; important books *Katz und Maus* 'Cat and Mouse', *Hundejahre* 'Dog Years', *Örtlich betäubt* 'Local Anaesthetic', *Der Butt* 'The Flounder', *Das Treffen in Telgte* 'The Meeting at Telgte', *Die Ratten* 'The Rats', *Unkenrufe* 'The Call of the Toad', and *Ein weites Feld* 'A Broad Field'.

1961-87: <u>Vidiadhar Surajprasad Naipaul</u>, Trinidad-born British novelist and essayist, became known for his works including *A House for Mr Biswas*; *The Loss of El Dorado*; *In a Free State*; *A Bend in the River*; and *The Enigma of Arrival*.

1962-97: <u>Phyllis Dorothy James</u>, English detective story writer, was author of detective stories such as *Cover Her Face*; series *A Mind to Murder*, *The Black Tower*, and *Death of an Expert Witness*; followed by *A Taste for Death*, *Devices and Desires*, and *The Skull Beneath the Skin*; and finally *The Children of Men*, *Original Sin*, and *A Certain Justice*.

1962-2006: <u>Jorge Mario Pedro Vargas Llosa</u>, Peruvian writer and essayist, became known for his novels *The Time of the Hero*, and *The Green House*; then wrote *Conversation in the Cathedral*, and *Aunt Julia and the Scriptwriter*; another novel *The War of the End of the World*; as well as *Aunt Julia and the Scriptwriter*; *The Feast of the Goat*; and *The Bad Girl*; he being recognized for 'his cartography of structures of power and his trenchant images of the individual's resistance, revolt, and defeat'.

1963-2004: <u>Jean-Marie Gustave Le Clézio</u>, French-Mauritian novelist, short story writer and essayist, published novels including *Le Procès-Verbal* 'The Interrogation', *Désert*, and *Onitsha*; as well as works such as *Le Déluge* 'The Flood'; *Poisson d'or*; *Terra Armata*; *Mondo and Other Stories*; and *The African*; he was recognized for his 'new departures, poetic adventure and sensual ecstasy', and as 'explorer of a humanity beyond and below the reigning civilization'.

1964-98: <u>Tom Stoppard</u>, British dramatist, produced radio plays such as *The Dissolution of Dominc Boot*; and works of *Rosencrantz and Guildenstern are Dead, Jumpers, Travesties, The Real Inspector Hound, Professional Foul, The Real Thing, Arcadia, Indian Ink*, and *The Invention of Love*; *The Russia House*; and especially *Shakespeare in Love*.

1964-2013: <u>Kenzaburō Ōe</u>, Japanese novelist, short story writer and essayist, became known for his books including *Aghwee the Sky Monster*; *A Personal Matter*; *The Silent Cry*; *Teach Us to Outgrow Our Madness*; *Rouse Up O Young Men of the New Age!*; and *In Late Style*.

1965-80: <u>Tsegaye Gabre-Medhin</u>, Ethiopian poet, playwright and essayist, wrote the acclaimed play *Tewodros*, another play *Petros at the Hour*, drama *The Oda Oak Oracle*, the poem *Zero Cataract* 'Blue Nile Falls', and dramas *Collision of Altars*.

1965-2008: <u>Grigore Vieru</u>, Moldovan poet, published the book *Poetry for Readers of All Ages*; poems *Ars Poetica*; *Friday's Star*; *I Do Not Hate You, Death*; *Autobiographical*; *Testament (To Mihai Eminescu)*; *In Your Language*; *Bliss*; and *A Letter from Bessarabia*; and also songs including *Melancholia*, becoming a representative of new literary movement in Moldova.

1966-2010: <u>Seamus Justin Heaney</u>, Irish poet and playwright, produced *Eleven Poems*; the collections *Death of a Naturalist*; *Wintering Out*; *North*; *Bog Poems*; *Stations*; *Preoccupations*; *The Spirit Level*; and *Electric Light*; as well as prose poems *Field Work*; *Station Island*; *The Haw Lantern*; and *Seeing Things*; also poems *District and Circle*; and *Human Chain*.

1967-2004: <u>Gabriel (José de la Concordia) García Márquez</u>, Columbian novelist and short story writer, was author of a series of books such as the novels *One Hundred Years of Solitude*; *The Autumn of the Patriarch*; *Chronicle of a Death Foretold*; and *Love in the Time of Cholera*; as well as the autobiographical *Memories of My Melancholy Whores*.

1968-2012: <u>Alice Ann Munro</u>, Canadian short story writer, became known for works such as *Dance of the Happy Shades*; *Lives of Girls and Women*; *Who*

Do You Think You Are?; *The Bear Came Over the Mountain*; as well as *The Love of a Good Woman*; *Hateship, Friendship, Courtship, Loveship, Marriage*; *Too Much Happiness*; and *Dear Life*; altogether, she is considered a 'master of the contemporary short story'.

1970-79: José Hernández, Argentinean poet, produced the epic poem *Martín Fierro*, consisting of *El gaucho Martín Fierro* 'The Gaucho Martín Fierro', and *La vuelta de Martín Fierro* 'The Return of Martín Fierro'.

1970-92: Toni Morrison (Chloe Anthony), US novelist, wrote the story *The Bluest Eye*; novels *Sula*, *Song of Solomon*, *Tar Baby*, *Beloved*, *Jazz* and *Paradise*; as well as the study *Playing in the Dark: Whiteness and the Literary Imagination*.

1972-82: Nichita Stanescu, Romanian poet, produced sensitive and outstanding works such as *The Book of Re-Reading*, *Epica Magna*, *The Imperfect Works/Creations*, and *Breathings/Respirations*.

1972-2004: Tomas Gösta Tranströmer, Swedish poet and psychologist, wrote a series of books including *Windows & Stones*; *Baltics*; *For the Living and the Dead*; *The Sorrow Gondola*, *The half-finished heaven*; and *The Great Enigma*; being characterized as 'through his condensed, translucent images, he gives us fresh access to reality'.

1972-2005: Lynda La Plante, English actress and writer, became a well-known author by her *Prime Suspect*, *Civvies*, *Framed*, *Comics*, *She's Out*, *The Governor*; then *Supply and Demand*, *Trial and Retribution*, *Killer Net*, *Profiler*, *Above Suspicion*; and novels such as *Bela Mafia*; many of them being adapted for television.

1975-91: Ioan Petru Culianu, Romanian linguist and philosopher, was professor the History of Religion, and published several outstanding books including *Expérience de l'Extase*, and *Eros et Magie à la Renaissance*, as well as *Dictionaire des Religions*, and *Out-of this World*.

1975-2003: Elfriede Jelinek, Austrian playwright, novelist and poet, wrote remarkable works such as the novels *Lust*, and *Greed*; and the plays *Das Schweigen* 'Silence', *In den Alpen* 'In the Alps', and *Das Werk* 'The Works'.

1976-2002: Akinwande Oluwole (Wole) Soyinka, Nigerian playwright and poet, became known for his books including *Ogun Abibiman*; *Ake*; *The Beatification of the Area Boy*; *The Burden of Memory, the Muse of Forgiveness*; and *King Baabu*.

1977-95: Joseph Alexandrovich Brodsky, Russian-born US poet and essayist, produced the poetry collections *A Part of Speech*, *To Urania*, and *On Grief and Reason*; and also the essay collection *Less Than One: Selected essays*.

1977-2003: John Maxwell Coetzee, South African novelist and essayist, became known especially for his novels such as *In the Heart of the Country*; *Waiting for the Barbarians*; *Age of Iron*; *Disgrace*; and *Elizabeth Costello*.

1981-2005: Wisława Szymborska-Włodek, Polish poet and essayist, created works including *Sounds, Feelings, Thoughts*; *People on the Bridge*; *View with a Grain of Sand*; *Poems*; *Miracle Fair*; and *Colon*.

1982-2005: <u>José de Sousa Saramago</u>, Portuguese writer, became known for his novels and allegories such as *Baltasar and Blimunda*; *The Year of the Death of Ricardo Reis*; *The Stone Raft*; *The Gospel According to Jesus Christ*; *Blindness*; and *Death with Interruptions*.

1983-2012: <u>Gao Xingjian</u>, Chinese author, wrote plays including *Bus Stop*, and *The Other Shore*; drama *Wild Men 'Savages'*; novel *Soul Mountain*, searching for roots in a shattered Communist China; fiction novel *One Man's Bible*; short stories *Buying a Fishing Rod for My Grandfather*; study *The Case for Literature*; and essays *Aesthetics and Creation*.

1985-2010: <u>Herta Müller</u>, Transylvanian German writer, was focused on German minority in the Communist Romania under repressive Nicolae Ceausescu's regime and also on modern history of Germans in Banat and Transylvania, and produced acclaimed novel *Atemschaukel* 'The Hunger Angel' depicting deportation of Romania's German minority to Stalinist Soviet Gulags during the Soviet occupation; and other works including *Drückender Tango* 'Oppressive Tango', *Der Teufel sitzt im Spiegel* 'The Devil is Sitting in the Mirror', and *In der Falle* 'In a Trap'.

1987-2005: <u>Mo Yan</u> (Guan Moye), Chinese novelist and short-story writer, published the novel *Red Sorghum Clan*, essays *The Wall Can Sing*, novellas *The Woman with Flowers*; and also works such as *The Garlic Ballads*; *Big Breasts and Wide Hips*; *The Republic of Wine*; *Pow!*; and *Life and Death Are Wearing Me Out*; being recognized for his 'hallucinatory realism' merging 'folk tales, history and the contemporary'.

1988-2004: <u>Imre Kertész</u>, Hungarian novelist, became known for his works including *The Failure*; *The Holocaust as Culture: Three Lectures*; *Kaddish for a Child Not Born*; and *Fatelessness*.

1990-2004: <u>Ferit Orthan Pamuk</u>, Turkish novelist and screenwriter, produced a series of works such as *The White Castle*, *The Black Book*, *The New Life*, *My Name is Red*, and *Snow*.

2000: <u>Roger Sell</u>, English linguist, literary analyst and professor of literary communication at Åbo Akademi University in Turku, Finland, wrote the book entitled *Literature as Communication: The Foundations of Mediating Criticism*, John Benjamins Publishing Company.

2000-08: <u>Linh Dinh</u>, Vietnamese writer, was author of the two collections of stories *Fake House*, and *Blood and Soup*; books of poems *All Around What Empties Out*, *American Tatts*, *Borderless Bodies*, and *Jam Alerts*; as well as the novel *Love Like Hate*.

2011: <u>Marjorie Garber</u>, US literary scholar, thoroughly studied and re-evaluated the use of term 'literature', as presented in her published work *The Use and Abuse of Literature*, Pantheon Books.

7. Communication technology

7.1. Roots of communication technology

Technology (from Greek τέχνη 'skill, art', and λογία 'study' from λογος 'account, explanation, speech, story') generally refers to making, modification, usage, and knowledge of tools, machines, techniques, crafts, systems, and methods of organization, in order to solve problems, improve pre-existing solutions of problems, achieve goals, handle applied input/output relations, or perform specific functions.

As a branch of technology, *information and communication technology* is related to the role of *unified communications* and the integration of *telecommunications* (e.g. telephone lines and wireless signals), *computers*, as well as necessary enterprise *software*, *middleware*, *storage*, and *audio-visual systems*, which enable users to access, store, transmit, and manipulate information.

Relatively long-distance communication was practiced from many thousands of years ago; and messages sent by men, horses, pigeons, fires, ships, flashing mirrors, etc. were frequently used in ancient and mediaeval times. For example, communication through *drawings* was practiced by indigenous tribes before 3500 BC; *metal mirrors* were used in AD 26-37 to reflect sunlight for sending messages to the ships during the Roman Emperor Tiberius' retirement to the island of Capri (Tiberius JCA, AD 14-37); and *firing cannon* and *raising flags* were practiced for signalling from a ship to other during the first circumnavigation (F Magellan and JS del Cano, 1519-22).

However, the long-distance communication was not a large scale technology until about four centuries ago, when new means of communication were developed on the basis of experience, inventive abilities, and empiric rather than scientific experiments, in order to satisfy the increasing needs to change information within the evolving societies.

Between the early 17th century and the late 19th century, information and communication technology developed due to a series of technical inventions and devices as well as scientific discoveries and theories, as follows:
- *Telescopes* (H Lippershey, 1608-10; I Newton, 1668);
- *Wave theory of light* (C Huygens, 1670-78; T Young, 1794-1820);
- *Heliometer* (P Bouguer, 1735-48);
- *Electricity* (B Franklin, 1746-48; AGA Volta, 1775-87; CA Coulomb, 1785; L Galvani, 1790-95; GS Ohm, 1827);

- *Electrolysis* (W Nicholson, 1800; M Faraday, 1831-34; AC Becquerel, 1835-37; WE Weber and FWG Kohlrausch, 1870-80);
- *Thermoelectricity* (TJ Seebeck, 1806-22; JP Joule, 1840-45);
- *Thermoconduction* and *signal frequency analysis* (JBJ Fourier, 1810-22);
- *Light diffraction* and *refraction* (J von Fraunhofer, 1814-21);
- *Light polarization* (AJ Fresnel, 1818-21);
- *Magnetic effect of electric current* (HC Oersted, 1820-25);
- *Electrodynamics* (AM Ampère, 1822-30);
- *Efficiency in conversion of energy* (S Carnot, 1824);
- *Magnetic field* (JB Biot and F Savart, 1828-30);
- *Frequency change in moving sources* (CJ Doppler, 1842; AHL Fizeau, 1845-49);
- *Electric bulb* and *printing photographs* (JW Swan, 1848-79);
- *Electroplating* (A Parkes, 1850-55);
- *Laws of diffusion* (AE Fick, 1855-80);
- *Spectrum analysis* (GR Kirchhoff and RW Bunsen, 1859);
- *Mathematical analysis of light waves* (LV Lorenz, 1863-69);
- *Diffraction theory of optical imaging* (E Abbe, 1865-90).

In the above stage, the development of information and communication technology could be briefly marked by achievements such as:

1608: *Refracting telescope* (from Greek τῆλε 'far' and σκοπεῖν 'to look or see') – first invented by aligning a *concave eyepiece* with a *convex objective lens* (H Lippershey, 1608-10);

1609: *News sheets* – first published on a *regular basis* at   Augsburg in Bavaria, Germany, and Strasbourg in   Alsace, France, keeping their inhabitants regularly informed;

1633-35: *Postal communication* – improved by placing boatmen under contract to make *regular crossings* with the mail between Dover and Calais, and then inland by keeping fresh horses and couriers in readiness at all times at fixed 'post stages' to carry *private mail* according to a published scale of payment (T Withering, 1632-35);

1668: *Reflecting telescope* – built by using a spherically ground metal *primary mirror* and a small *diagonal mirror* in such an optical configuration to avoid chromatic aberration (I Newton, 1667-68);

1692-1704: *Ricochet firing* – first introduced for communication along *trenches* and between *tower-bastions* in the French systems of fortification (SP de Vauban, 1669-1704);

1782: _Printed code for signalling with flags_ – used by the British fleet during its _manoeuvres_ and military _operations_ (R Howe, 1778-82);

1792: _Semaphore_ (from Greek σήμα 'sign' and φωρος 'bearer') – first included into a _semaphore line_ as a _practical telecommunication system_, by which messages were successfully sent between Paris and Lille (C Chappe, 1791-94);

1794: _Non-electric telegraph_ (from Greek τηλε 'far' and γραφειν 'write') – a visual system using semaphore and flag-based alphabet, which depended on a line of sight for communication (C Chappe, 1791-94);

1799-1805: _Signal-Book for the Ships of War_ – included the _flag signal_, _numerical code_ and the _listed meanings_, enabling British commanders to send orders to the ships over relatively great distances; it was used by English admiral Horatio Nelson at _Trafalgar_ to fly from his masts the message 'England expects that every man will do his   duty' (H Edles, 1799);

1809: _Electric telegraph_ – first invented as a _system of communication_ using 35 wires with gold _electrodes in water_ and at the receiving end 2000 feet the message was read by the amount of gas caused by electrolysis (ST von Sömmerring, 1809-23);

1821: _Heliograph_ (from Greek Ηλιος 'sun' and γραφειν 'write') – a wireless telegraph that signals by _flashes of sunlight_ reflected by a mirror, as a simple but effective instrument for instantaneous optical communication over long distances, initially called 'heliotrope' (JCF Gauss, 1821-43);

1825: _Elecromagnet_ (from Greek ήλεκτρον via Latin _electrum_ 'amber', and from Greek Μαγνήτις λιθος 'stone from Magnesia', i.e. from the ancient city of Asia Minor) – a type of _magnet_ in which the magnetic field is produced by _electric current_; the invention of electromagnet laid the foundations for a large scale development in electronic communications, such as relays, loudspeakers, hard disks and magnetic resonance imaging (W Sturgeon, 1824-36);

1828: _Electrolytic telegraph line_ – first applied at a race track in Long Island, the USA, using a _bare electrical wire_ to transmit _sparks_ generated by electric current, which were recorded on a ribbon of moistened litmus paper to burn _dots_ and _dashes_ on a spool mechanically revolved (HG Dyar, 1826-28);

1830-35: _Communication ending by a bell to strike_; and _electric relay_ – practical devices developed by using the electromagnet (J Henry, 1829-35);

1831-37: _Single-wire telegraph system_ – for using a practical _electric telegraph_ with commercial success; _first long-distance electric telephone line_ – two miles of wire for sending messages at Speedwell Ironworks near Morristown, New Jersey, USA; and _Morse code_ – a series of _on-off tones_, _lights_, or _clicks_ understandable by a skilled listener or observer without special equipment (SFB Morse and AL Vail, 1831-44);

1845: _Electrodynamometer_ – an instrument for measuring electric currents by the _attraction_ or _repulsion_ between current-bearing coils, that from its early stage was provided with _mirror_ and _scale_ for reading deflection (CF Gauss and WE Weber, 1843-45);

1849: _Nova Scotia Pony Express_ – an innovative combination of technologies including _(i)_ regular _Cunard steam pockets_ between Liverpool (England) and Halifax (Nova Scotia, Canada), linked to fast _charter steamboats_ on the Bay of Fundy (Canada); _(ii) telegraph_ available from Saint John (New Brunswick, Canada) to New York (USA); and _(iii) riders_ on fine _horses_ available for dispatches on 144 miles from Halifax to Victoria Beach (Manitoba, Canada) in an average of 8 hours; altogether forwarding European news to be conveyed by chartered steamship and telegraphed to the New York newspapers (Associated Press);

1863: _Telegraphy equipment_ – a business known as 'Siemens Brothers' providing _signalling_ and _communication devices_, such as power transmission lines, carrier transmission equipment, and electrical railway signals (EW von Siemens, 1847-67);

1871: _Electric generator_ – a device that converts mechanical energy to electrical energy, commercially used for _electroplating_ and _electric lighting_ (ZT Gramme, 1869-71);

1873: _Orthochromatic film_ – a photographic film sensitive to green, blue and violet light, consisting of a base of celluloid covered with a photographic emulsion and used to make _negatives_ or _transparencies_; and _photometer_ – an instrument for measuring the brightness of objects, including the _photographic plate_ (HW Vogel, 1870-73).

7.2. High technology communication

High technology is the technology that uses highly sophisticated equipment and advanced engineering techniques, such as electrical engineering, automotive, nuclear physics, photonics, nanotechnology, telecommunications, information technology and systems, biotechnology, robotics, artificial intelligence, etc.

An important role in the development of high technology was played by *telecommunication* (from Greek *τηλε* 'distant' and Latin *communicare* 'to share') as the communication at distance using technological means, particularly through electrical signals or electromagnetic waves, including telegraph, telephone, radio, television, video-telephony, satellite, digital cinema, computer networks, the Internet, and so on.

Along with wave-particle duality, the electromagnetism, radiation, quantum mechanics, and other scientific branches; the high technology recorded a rapid evolution and diversification of its fields of study, including:

(i) Thermal radiation, cosmic microwave radiation, electromagnetic radiation, black-body radiation, and thermionics;

(ii) Radioactivity, radioelements, radioisotopy, radio-astronomy, radiochemistry, radiobiology, radiography, radiology, radioscopy, radiotherapy, and radioimmunology;

(iii) Radiotelegraphy, radiotelephony, radio-communication, and radio-engineering;

(iv) Nuclear physics, nuclear collision, nuclear transmutation, nuclear fission, nuclear fusion, spectrology, spectroscopy, quasars, pulsars, string, and neutrinos;

(v) Particle distribution, particle statistics, particle disintegration, particle acceleration, quarks, anti-particles, light shifting, photometry, superfluidity, superconductivity, cinematography, cybernetics, and television;

(vi) Radar and sonar techniques, masers, lasers, computers, transistors, microprocessors, teleprocessors, telesoftware, and integrated circuits.

The emergence of high technology was preceded by a series of scientific discoveries; especially those concerning the fundamental properties of waves, particles, materials; and the generation, transmission and reception of signals, as conveyors of energy and information. A theoretical approach of signalling process at a certain frequency band was first possible by the *Fourier transform*

$$\Phi(v) = \int_{-\infty}^{+\infty} \varphi(t) \cdot e^{-2\pi \cdot i \cdot t \cdot v} \cdot dt$$

of a integrable function of time $\varphi(t)$, that, for a rectangular pulse $\varphi(t) = 1/(2t')$ in a finite interval of time $-t' < t < +t'$, and $\varphi(t) = 0$ outside the interval, becomes

$$\Phi(f) = [1/(2t')] \cdot \int_{-t'}^{+t'} e^{-i \cdot f \cdot t} \cdot dt = [1/(2t')] \cdot sin(f \cdot t'),$$

where v is the signal frequency, and $f = 2\pi \cdot v$. This transform shows that $\Phi(0) \rightarrow 1$ when $f \cdot t' \rightarrow 0$, and $\Phi(f) = 0$ when $f = \pm\pi/t', \pm 2\pi/t',...$, whilst the amplitude of the carrier decreases when f increases and finally vanishes. In the time interval $2t'$, first annulment takes place when $f = \pi/t'$ and therefore the most energy of signal corresponds to frequencies less than π/t', so that the frequency band can be roughly estimated as $\Delta\varphi \sim 1/(2t')$. As the signal duration is $\Delta t = 2t'$, and then $\Delta\varphi \cdot \Delta t \sim 1$; it follows that a longer signal is associated with a shorter pulse spectrum, and in reverse. More rigorously, this relationship can be expressed by the inequality $\Delta f \cdot \Delta t \geq 1$ (JBJ Fourier, 1810-22), and then used to show that the persistence of the information conveyed by a signal is mainly because of its reduced frequency band or spectrum.

In information and communication, like in other fields of interest or activity, the high technology had not a scientific support until the first *colour photography* (T Sutton, 1861), the *theory of electromagnetism* (JC Maxwell, 1861-73), the concept of *entropy* (RJE Clausius, 1865), and the *principle of increasing entropy* (L Boltzmann, 1868-77) became understood and available for technical purposes. Based on the second law of thermodynamics in connection with the dissipation of energy, the *entropy* (Greek ἐντροπία 'a turning towards' coming from ἐν 'in' and τροπή 'a turning') is a quantity showing the state of disorder developed within a system during the process of conveying energy, information, or other utilities, and can be generally expressed as

$$S = -k \cdot \Sigma_{i=1}^{n} p_i \cdot ln(p_i),$$

where S is the entropy, k is a constant such as Boltzmann's constant [$1.3805 \cdot 10^{-23}$ J/K], i is a number indicating a single microstate, n is the total number of microstates, and p_i is the probability of i-microstate.

The generation, transmission, processing and reception of information are also dissipative transformations diminishing the differences in energy density between their subsystems and surroundings. *Information entropy* is defined as: a measure of the uncertainty associated with a random variable; a measure of the average information content missed when the value of the random variable is not known; or the amount of information contained per average instance of a character in a stream of characters.

The emergence and advance of high technology were based on theoretical and practical discoveries, inventions, methods, systems, devices, theories and principles, such as: *Boolean algebra* (G Boole, 1847-54); *refractive index* depending on *polarizability* (LV Lorenz and HA Lorentz, 1863-78); *canal rays* 'Kanalstrahlen' (E Goldstein, 1876-86); *three-phase alternating-current generator* (E Thomson, 1880-92); *rollfilm* and *photographic camera* (G Eastman and TA Edison, 1880-92); *long-range transmission* (W Stanley, 1885); *high-frequency coil* and *air-core transformer* (N Tesla, 1885-88); *radio waves* 'Hertzian waves' (HR Hertz, 1887); *electromagnetic rays* 'X-rays' (WK von Röntgen, 1895); *wireless transmission* (E Rutherford, 1895-96); *broadening of spectrum emission lines* (P Zeeman, 1896); *cathode-ray oscilloscope* (F Braun, 1897); *optic effects of positive (canal) rays* (JJ Thomson and WCW Wien, 1897-1911); *quantum theory* (MKE Planck (1898-1900); *diode valve* (JA Fleming, 1904); *amplitude modulation* and *radio broadcast* (RA Fessenden, 1906).

A breakthrough in physics and implicitly in high technology arouse from the *wave-particle duality* (L-V de Broglie, 1923-28) whereby the propagation of electromagnetic waves with light speed c and, by virtue of their motion, associated with particles of relativistic momentum as pulse p [kilogram-metre per second], the wavelength $\lambda = c/f$ [metre] is given by the simple equation

$$\lambda = h/p,$$

where h is Planck's constant [$6.626 \cdot 10^{-34}$ joule-second]. The wave propagation is related not only to photons but also to electrons and other fundamental particles, including neutrinos and antineutrinos which are unrestrictedly crossing the entire universe. As asserted by the wave-particle duality, the larger particles and bodies are also associated with waves, but their lengths are too short, i.e. frequencies too high, to be detected. Notably, the more harmonic the more persistent the waves are.

A more comprehensive understanding of wave propagation was possible due to the development of *wave mechanics* (E Schrödinger, 1926-30) and *quantum mechanics* (PAM Dirac, 1928-30). In the context of signal processing, *Dirac delta function*, or unit impulse function (that is zero everywhere except at the single point of origin), is regarded as a kind of weak limit in a sequence of functions having a tall spike at the signal origin, and making it extremely useful in many natural processes involved in the high technology communication.

Further advance in information and communication technology consisted of achievements including: *transmission theory* and *regeneration theory* (H SyQuest, 1928); *transmission of information* and *storage principle* (RVL Hartley, 1928-36); practical *television camera* and *colour receiver* (VK Zworykin, 1928-39; EFW Alexanderson, 1930-55).

In this stage, the *cybernetics* (Greek κυβερνήτης 'steersman, governor' from the same root as κυβερνητική 'government') emerged as a comparative study of automatic communication and control in functions of living bodies and in mechanical electronic systems (N Wiener, 1932-48). Cybernetics attempts to model the feedback and control mechanisms, which are inherent in *intelligent behaviour*; in *computer science*; and in *artificial intelligence* as an intelligent function, regardless of the particular way it can be achieved.

Computer science directly applies the concepts of cybernetics to the control of devices and the analysis of information, such as *design patterns, robotics, decision support system, cellular automata*, and *simulation*.

Other new technologies, principles and theories consisted of *differential analyser* or *analogue computer* (DR Hartree, 1932-58); *electrostatic xerography* (CF Carlson, 1937-38); *robotics* (I Asimov, 1941-50); *electronic calculating machine* (JV Atanasoff and C Berry, 1942); *electronic computer* and *stochastic computing* (J von Neumann, 1945-53); *point-contact transistor* (WH Brattain *et al.*, 1947); *holography* (D Gabor, 1947); *sampling theorem* and *information theory* (CE Shannon, 1945-49).

According to the fundamental *storage principle* (RVL Hartley, 1928-36; CE Shannon, 1945-49), the storable information in a system is proportional to the logarithm log_bN of the number N of possible states of that system. Choosing the base b of logarithm as the number 2, the unit measuring information, called *shannon*, is equal to the information content of one *bit* (*b*inary dig*it*) as the capacity of a system which can exist in only *two states*, and the smallest unit of information in computers and communication theory. Therefore, a system with N possible states has $log_2N = n$ bits, whence $N = 2^n$.

A character of text depends on hardware architecture of the computer, but it is almost always encoded using an octet called *byte* = 8 *bits* = 2^8 = 256 distinct values (e.g. integers). In entropy content, such units measuring information would be used taking into account that: 1st, the amount of entropy is not always an integer number of bits; and 2nd, many data bits may not convey information, e.g. data structure often store information redundantly, or have identical sections regardless of the information in the data structure. This *binary logic*, also called *Boolean computing*, was essential for developing the modern technology of information and communication.

In its later stage, the high technology related to information and communication progressed faster than before as follows: *maser* (CH Townes, 1951-53); *optical pumping* (A Kastler, 1952-65); *ammonia maser* (CH Townes and J Gordon, 1953); *large electron-positron collider* (CERN, 1954-89); *laser* (NG Basov and AM Prokhorov, 1955); *miniature television set* and *personal computers* (CM Sinclair, 1958-70); *cyborg* '*cyb*ernetic *org*anism' (M Clynes and N Kline, 1960; DS Halacy, 1965); *datagram* '*data* tele*gram*' and *RUNCOM* '*RUN COM*mands' (L Pouzin, 1963-64); *ARPANET* '*A*dvanced *R*esearch *P*roject *A*gency *NET*work' (R Kahn and V Cerf, 1969); the precursor of *Internet* (L Kleinrock and DC Engelbart, 1969-71); computer *mouse* '*m*anually *o*perated *u*ser *s*election equipment' and *e-mail* (DC Engelbart, 1970-89); *microprocessor* 'Intel 4004' (F Faggin *et al.*, 1971); *PUMA* '*p*rogrammable *u*niversal *m*anipulation *a*rm' (V Scheinman, 1975-77); *Windows 2000* (B Gates, 1982-2000); *integrated circuit* (J St Clair Kilby, 1987-95); *semiconductor heterostructures* (ZI Alferov and H Kroemer, 1990-98); *CCD* '*c*harge-*c*oupled *d*evice' *sensor* (WS Boyle and GE Smith, 1996-2008); *cell-probe lower bounds* (M Patrascu *et al.*, 2006-11).

The advance in high technology communication is selectively and chronologically displayed below.

1876: *First practical telephone* (from Greek *τῆλε* 'far' and *φωνή* 'voice')
‒ a telecommunication device that permits two or more users to conduct a *conversation* when they are not   in the same vicinity of each other to be heard directly (AG Bell, 1875-77).

1877: *Phonograph* (from Greek *φωνή* 'sound, voice' and *γραφειν* 'write') or *gramophone* (from Greek *γράμμα* 'letter' and *φωνή* 'voice') ‒ a device introduced for reproducing *sound recordings*, initially on *tinfoil* around a grooved cylinder (TA Edison, 1877-91).

1879: *Incandescent light bulb* ‒ an electric light produced with a *filament wire* heated to a high temperature by an electric current passing through, until it glows (JW Swan, 1848-79; TA Edison, 1877-91).

1889: *Direct dial telephone* ‒ a telephone designed to make *calls without connection* by an operator (AB Strowger, 1889-91).

1895: _Cinematograph_ – a machine used to project _images on a screen_, by which the first _film_ was produced (AMLN Lumière and LJ Lumière, 1895-1903).

1901: _Wireless telegraphy_ – a system of long-distance radio-transmission by which first _transatlantic radio signals_ were sent from Cornwall, England, to Newfoundland, Canada (G Marconi, 1895-1901).

1920: _Radio broadcasting_ – the distribution of audio content to a dispersed audience, usually by _electromagnetic waves_, which was first commercially performed by a radio station KDKA licensed in Pittsburgh, Pennsylvania, USA, for a power of 50 kilowatts on a 1020 kilohertz frequency (Westinghouse Electric Corporation).

1925: _Television_ (TV) – a telecommunication medium used to transmit and receive moving _images_ associated by _sound_, which was first publicly demonstrated as a black-and-white system, and then as a purely _electronic colour_ system (JL Baird, 1924-26).

1942: _Frequency hopping spread spectrum_ – a technique for _synchronizing random frequencies_, with a receiver and transmitter, which became widely used in _telecommunications_, and paved the way for wireless communications via _Internet_ and _Wi-Fi_ '_Wi_reless-_Fi_delity' (H Lamarr and G Antheil, 1941-42).

1947: _Cell phone_ – a cell-based approach leading to _cellular phones_ (D Ring and WR Young, 1947-48) \ First broadcast by a _full-scale commercial television_.

1949-71: _Electronic book_ 'e-book' – a _book_-length publication in _digital form_, consisting of text, images, or both, readable on _computers_ or other _electronic devices_ (R Busa, 1949; ÁR Robles, 1949; A van Dam, 1965-68; MS Hart, 1971).

1963: _Geosynchronous communication satellite_ – a _satellite_ in _geosynchronous_ orbit, with an orbital period the same as the Earth's rotation period, which enabled the world's first _satellite-relayed telephone call_ (H Rosen, 1962-63).

1963-64: _Telecommunication by fiber optics_ – a silica-based _optical waveguides_ used to transmit light via _total internal reflection_ (CK Kao, 1963-81).

1964: _Personal computer_ (PC) – a general-purpose computer useful for individuals by its size, capabilities and sale price, which is intended to be operated directly by and _end-user_, without intervening computer operator, such as *desktop computer*, *laptop*, *Netbook*, *tablet*, or *handheld PC* (PG Perotto, 1962-64).

1971: *Computerized telephone switching* – a modern technology of telecommunication that prevents overloads by monitoring call centre traffic and prioritizing tasks on *phone switching systems* to avoid perturbations during peak calling times (ES Hoover, 1967-71).

1971-72: *Electronic mail* (email) – a method to exchange *digital messages* from a sender to one or more receivers, which operates across the *Internet* or other *computer networks* (RS Tomlinson, 1971-73).

1978: *TeX computer typesetting* – a system designed to allow anybody to produce *high-quality books* with reasonably minimal effort and the same results on all computers (DE Knuth, 1976-78).

1989: *Global positioning satellite* (GPS) *system* – comprehensive worldwide navigational system, previously used for ship navigation and military purposes, became first commercially handheld GPS receiver, and then was developed as a network of 24 satellites placed into orbit by the US Department of Defence; GPS devices having capabilities such as maps displayed in human readable format via text or in a graphical format, turn-by-turn navigation directions, directions fed directly to an autonomous vehicle including robotic probe, traffic congestion maps associated with suggested alternative directions, and information on amenities.

1990: *World Wide Web* (www) – a system of interlinked *hypertext* documents accessed via *Internet*, by a *web browser* for viewing *web pages* that may contain text, images, videos and other multimedia, and navigating between them via hyperlinks (T Berners-Lee and R Cailliau, 1989-90, at CERN, 1954-89).

1992: *Text messaging* – the act of composing and sending a brief *electronic message* between two or more mobile phones, or fixed or portable devices over a *phone network* (N Papworth, 1991-92).

1993: *Internet radio broadcasting* – an *audio service* transmitted via the *Internet* (C Malamud, 1993-97).

1994-99: *Blog* (web log) – in modern version evolving from the *online diary*, as a discussion or informal site published on the www, which consists of discrete *entries* 'posts', usually displayed in reverse chronological order; e.g. *personal blogs, microblogging, corporate and organizational blogs* (J Hall, 1994; D Winer, 1999).

1999: *Sirius Satellite Radio* – a *satellite radio service* operating in North America, owned by *Sirius XM Radio* (MA Rothblatt *et al.*, 1999-2002).

2002: *Linkedin* – a business-oriented '*Linked In*' *social networking service* that is mainly used for *professional networking* such as business connections and job searching (RG Hoffman and A Blue, 2002-03).

2003: *Myspace* – a *social networking service*, especially for music emphasis, owned by Specific Media *LLC* '*Limited Liability Company*' (C DeWolfe and T Anderson, 2003-04).

2004: *Facebook* – an online *social networking service* developed to create *personal profiles*, add other users, exchange messages, and receive automatic notifications for updating profiles (ME Zuckerberg *et al.*, 2003-06).

2005: *YouTube* – a *video-sharing* website, on which users can upload, view and share videos; the company is based in San Bruno, California, and via Adobe *Flash Video* and *HTML5* '*HyperText Markup Language* (version *5*)' technology displays a variety of *user-generated* and *corporate media* video content (S Chen *et al.*, 2005-06).

2006: *Twitter* – an online *social networking* and *microblogging* service, based in San Francisco, California, enabling users to send and read short *140-character* text messages, called *tweets*, which can be accessed through the website interface, *SMS* '*Short Message Service*', or mobile *application* device; the registered user can read and post tweets, while the unregistered ones can only read them (EC Williams *et al.*, 2006-07).

2007-14: *iPhone models* – a series of *smartphones*, including *iPhone 3G, 3GS, 4, 4S, 5, 5C*, and *5S*, and ranging from 8 to 64 gigabytes; iPhone has Wi-Fi, is connectable to many cellular networks, and can shoot video, take photos, play music, send and receive email, brose the web, send texts, cope with GPS navigation, record notes, do mathematical calculations, and receive visual voicemail (all models are designed and marketed by Apple Inc.).

Computer-mediated communication has been traditionally referred to those communications which take place via computer-mediated formats, e.g. instant messaging, email, and chat rooms; but today is also applied to other forms of text-based interaction such as text messaging. As many recent studies involve Internet-based *social networking* supported by *social software*, the computer-mediated communication is

229

defined as any communication that take place through the use of two or more electronic devices. Therefore, this kind of communication became valuable on providing a better communication and better first impressions (J Walther, 1992-96). Further studies indicated that computer-mediated communication allows more closeness and attraction between two individuals than a face-to-face communication (A Ramirez and S Zhang, 2007). Providing opportunities for language learners to practice their language, the computer-mediated communication is widely discussed in language learning, for example by use of email or discussion boards in different language classes, and is claimed that information and communications technology 'bridge the historic divide between speech...and writing' (M Warschauer, 1999-2006).

Communicative transaction occurs through the use of two or more networked computers, and consists of fifteen subcategories: *(i) bulletin board systems, (ii) communication software, (iii) email, (iv) file sharing, (v) Internet forums, (vi) instant messaging, (vii) online chat, (viii) social media, (ix) social networking services, (x) telecommuting, (xi) teleconferencing, (xii) telepresence, (xiii) unified communication, (xiv) voice over Internet Protocol,* and *(xv) web conferencing.*

Communication by technological means became easier, faster and more frequently used. The advance of technology offers new possibilities of communication, but which of them will actually break through to remain a facet of daily life depends on our needs.

In the immediate future, the circulation of newspapers, magazines, and books would be lower than in present, and the print industry is expected to lose 25% of its jobs over the next few years. The decline of print industry will be because the Internet has and continues to shape the way of communication, and meanwhile applications such as eBook, iBook, and audiobook are expected to increase greatly. Also in a few years, newspapers, books, and advertisements will be mainly created and made for online, enhancing people into a global interaction.

Recent investigations indicate five theses about the future of *innovation communication* (F Bethge, 2012):

- Innovation development and innovation adaption will be synchronized,
- Content syndication will transform adaptation into participation,
- Half of the communication process about innovation will be delivered automatically,
- Face-to-face communication remains essential for building up innovator teams, and

- Online/Offline-Communities will transform participation into social integration.

At present, the high technology covers a wide range of branches, most of them directly or indirectly related to information and communication at global, national, local and personal levels, such as:

Information technology, or *Information and Communications Technology* (ITC) – the study, design, development, implementation, support, or management of computer-based information systems; the unified communications and integration of telecommunications (telephone lines and wireless signals), computers, and necessary enterprise software, middleware, storage and audio-visual systems, enabling users to access, store, transmit and manipulate information; and *telecommunication technology* comprising technological means, particularly through electrical signals or electromagnetic waves (telephone, radio, television, internet, local and wide area networks);

Spatial technology – referring to any software or hardware that interacts with real world location, especially at the Earth's surface;

Electronics – dealing with electrical circuits (analogue circuits and digital circuits) that involve active electrical components, including vacuum tubes, transistors, diodes and integrated circuits, as well as associated passive interconnection technologies;

Biotechnology – using living systems and organisms to develop or make products, or more precisely any technological application that uses biological systems, living organisms or derivatives thereof, to make or modify products or processes for specific purposes;

Process technologies – as the machines, equipment and devices that contribute to an operation transforming materials and information, and customers to add value; these technologies including a wide range of building blocks (products, units and skids) such as heat transfer, membrane filtration; cheese and butter distillation and evaporation; blending and mixing technologies designated to operate at maximum efficiency in modern processing lines within dairy, food, beverage, brewery, etc.; as well as their linkage in processing lines;

Low-temperature and *high-temperature technologies* – introducing innovative solutions to provide and improve materials and equipments for working either at very low or very high temperatures, for example by using wolfram, quartz glass, ceramic materials and graphite to the processes which take place at temperatures more than 1000°C;

Robotics – dealing with the design, construction, operation and application of robots, as well as computer systems for their control, sensory feedback and information processing;

Nuclear technology – involving reactions of atomic nuclei, notably nuclear power, nuclear medicine and nuclear weaponry, which found applications from smoke detectors and gun sights to nuclear reactors and sophisticated nuclear weapons; and also industrial and commercial applications, food processing and agriculture;

Medical high-technology – including da *Vinci-Si HD* Surgical Robotic System (providing enhanced capabilities, including a magnified, high-definition, 3D), *CyberKnife* (as a method of delivering radiotherapy used for targeting treatment more accurately than standard radiotherapy), *InnerCool RTx* Endovascular System (for cooling and warming that provides advanced whole body temperature modulation therapy in a closed loop system from the inside out), Olympus *VisiGlide* (a single use guide wire for making ductal navigation easier and faster), *PillCam* (endoscopic camera/capsule for healthcare market), Siemens *SOMATOM* Definition Flash (for scanning at Flash speed, so the diagnostic images can be acquired from almost any patient), Antenna *Pill* (electronic device as a bar code that ingested emits a signal warning the patient to take his/her medicine);

Genetic engineering – the direct manipulation of an organism's genome using biotechnology, by methods of cloning genetically engineered molecules in foreign cells, creation of transgenetic organisms, etc.;

Computing – any goal-oriented activity requiring, benefiting from, or creating computers, including computer engineering, software engineering, computer science, information systems and information technology;

Direct-digital control – by which a single digital computer replaces a group of single-loop analogue controllers, increasing the computational ability and permitting the application of more complex and advanced control techniques;

Nanotechnology – as manipulation of matter on an atomic, molecular, and supramolecular scale (matter with at least one dimension sized from 1 to 100 nanometres), until now using *45, 32, 22, 14*, and *10 nanometre* (referring to the average half-pitch of a memory cell) *chips* produced by Matsushita and Chartered Semiconductor, *AMD* '*A*dvanced *M*icro *D*evices' Inc., *IBM* '*I*nternational *B*usiness *M*achines' Corporation, Samsung, and Chinese *SMIC* '*S*emiconductor *M*anufacturing *I*nternational *C*orp;

232

Satellite imagery or *Satellite-Imaging technology* ⁻ consisting of images of the Earth or other planets collected by artificial satellites;

Artificial intelligence (AI) ⁻ the technology and a branch of computer science that studies and develops intelligent machines and software, with goals of deduction, reasoning, problem solving, knowledge representation, planning, learning, natural language processing, motion and manipulation, perception, social intelligence, creativity and general intelligence.

7.3. Advance modelling

According to data above, the technology in general and communication technology in particular advanced from around four centuries ago to present-day in stages approximately delimited such as:

\ AD 1600 → 415 BP \ *Refracting telescope, posting*
\ AD 1640 → 375 BP \ *Reflecting telescope*
\ AD 1680 → 335 BP \ *Ricochet firing*
\ AD 1720 → 295 BP \ *Heliometer, electricity*
\ AD 1760 → 255 BP \ *Signalling code, semaphore*
\ AD 1800 → 215 BP \ *Telegraph, electric relay*
\ AD 1840 → 175 BP \ *Electric generator, film*
\ AD 1875 → 140 BP \ *Telephone, cinema*
\ AD 1915 → 100 BP \ *Radio, TV, cell phone*
\ AD 1950 → 65 BP \ *PC, email, TeX, www*
\ AD 1990 → 25 BP \ *Internet, YouTube, Twitter, iPhone*

The transitional years delimiting these stages are converted in years from the origin of life c.4200 million years BP, as follows:

\ 4200 - 0.000415 = 4199.999585 \ 4200 - 0.000375 = 4199.999625 \
\ 4200 - 0.000335 = 4199.999665 \ 4200 - 0.000295 = 4199.999705 \
\ 4200 - 0.000255 = 4199.999745 \ 4200 - 0.000215 = 4199.999785 \
\ 4200 - 0.000175 = 4199.999825 \ 4200 - 0.000140 = 4199.999860 \
\ 4200 - 0.000100 = 4199.999900 \ 4200 - 0.000065 = 4199.999935 \
\ 4200 - 0.000025 = 4199.999975 \

Nevertheless, the advance of communication technology can be integrated into the life evolution on the basis of literature development (see table by the end of subchapter 6.3) with its latest $N/10^6 = \zeta$ sequences delimited by values of argument \1.880599\... \1.880600...\1880601\, which comprise shorter $N/10^7 = \eta$ sequences suitable for displaying the course of communication technology. As the final time was estimated as $t_\bullet = 4400$ million years from the origin of life, the transitional times $t_\eta = t_\bullet \cdot tanh(\eta)$, delimiting the sequences of communication technology, are calculated in million years ago as

$$t_{1.8805990} = t_\bullet \cdot tanh(1.8805990) \approx 4199.999586;$$
$$t_{1.8805991} = t_\bullet \cdot tanh(1.8805991) \approx 4199.999626;$$
$$t_{1.8805992} = t_\bullet \cdot tanh(1.8805992) \approx 4199.999665;$$
$$t_{1.8805993} = t_\bullet \cdot tanh(1.8805993) \approx 4199.999704;$$
$$t_{1.8805994} = t_\bullet \cdot tanh(1.8805994) \approx 4199.999743;$$
$$t_{1.8805995} = t_\bullet \cdot tanh(1.8805995) \approx 4199.999782;$$

$$t_{1.8805996} = t_\bullet \cdot tanh(1.8805996) \approx 4199.999821;$$
$$t_{1.8805997} = t_\bullet \cdot tanh(1.8805997) \approx 4199.999860;$$
$$t_{1.8805998} = t_\bullet \cdot tanh(1.8805998) \approx 4199.999899;$$
$$t_{1.8805999} = t_\bullet \cdot tanh(1.8805999) \approx 4199.999938;$$
$$t_{1.8806000} = t_\bullet \cdot tanh(1.8806000) \approx 4199.999977.$$

The timeline and sequences of communication technology are summarized in the table below.

$N/10^7$ $= \eta$	Time (million years)		Sequences of communication technology
	from origin $t_\eta = t_\bullet \cdot tanh(\eta)$	from present $t_\eta - 4200$	
	4400	+200	
...	...	...	
1.8806001	4200.000016	+0.000016 (AD2031)	_
1.8806000	4199.999977	-0.000023 (AD1992)	_Internet, YouTube, Twitter, iPhone
1.8805999	4199.999938	-0.000062 (AD1953)	_PC, email, TeX, www
1.8805998	4199.999899	-0.000101 (AD1914)	_Radio, TV, cell phone
1.8805997	4199.999860	-0.000140 (AD1875)	_Telephone, cinema
1.8805996	4199.999821	-0.000179 (AD1836)	_Electric generator, film
1.8805995	4199.999782	-0.000218 (AD1797)	_Telegraph, electric relay
1.8805994	4199.999743	-0.000257 (AD1758)	_ Signalling code, semaphore
1.8805993	4199.999704	-0.000296 (AD1719)	_Heliometer, electricity
1.8805992	4199.999665	-0.000335 (AD1680)	_Ricochet firing
1.8805991	4199.999626	-0.000374 (AD1641)	_Reflecting telescope
1.8805990	4199.999586	-0.000414 (AD1601)	_Refracting telescope, posting
...	...	...	
0	0	-4200	

In 2013-14, the South Korean multinational conglomerate corporation *LG* (*L*ucky-*G*oldstar) *Electronics* released the OLED (*O*rganic *L*ight-*E*mitting *D*iode) commercial *televisions with thinner screens*, more efficient and capable of displaying greater definition images than conventional *LCD* (*L*iquid *C*rystal *D*isplay) and plasma screens.

Meanwhile, there was recorded a major expansion in the use and capabilities of technologies such as *3D printing* and *autonomous cars*.

In addition, the Advanced Cell Technology announced that it created new *human embryonic stem cells* by fusing DNA from an adult with an enucleated egg cell, as a form of human cloning that open the way to generate healthy replacements of diseased or damaged cells in patients (A Park, 2014).

Also in 2014, *Beth Israel Deaconess Medical Center* (BIDMC) announced that an international team of neuroscientists and robotics engineers has demonstrated the viability of *direct brain-to-brain communication* in humans; the highly novel findings describing the successful transmission of information via the internet between the intact scalps of two human subjects located 5000 miles apart.

High technology advance from present to future

Running researches and experiments

(i) Remembering that the current computing is based on binary logic (zeroes and ones), also called Boolean computing, a new type of computing architecture is experimented for enabling to store information in the frequencies and phases of periodic signals which could work more like the human brain to compute using only a fraction of the energy of today's computers. At Penn State, the USA, a first device was created using a thin film of vanadium dioxide (VO_2) on a titanium dioxide (TiO_2) substratum, as a coupled system, to create an oscillating switch and to synchronize them for simulating the biological synchronization, thus providing the basis for *non-Boolean computing*, and creating a different kind of computing 'associative processing' which is an analogue rather than digital way to compute.

(ii) Experiments with cells of the fruit fly are carried out for developing a sophisticated computer modelling simulation to explore how *organic cells react to changes in the environment.*

(iii) A new *cutting-edge tablet with advanced vision capabilities* for mobile devices uses a 7-inch screen equipped with two black cameras, infrared depth sensors and advanced software, which can capture three-dimensional images and objects, and is developed as part of a Google Inc. research effort dubbed Project Tango.

(iv) Cox Communications Inc. became the biggest US cable operator to commit for rolling out a *gigabit-speed broadband* offering to all its residential customers.

Current prototypes

Human tissue printer - sophisticated and experimental '3D Bio-printer' capable of 'printing' arteries;

Advanced robots - 'Ecci', one of the most advanced robots in the world, that mimics bones and joints, and may be with ability to learn from its mistakes;

Thought controlled prosthetics - 'Darma', for controlling with mind robotic limbs, by implanting a chip in the brain;

Wireless power - transmission of electricity from a power source to an electrical load without man-made conductors, such as using direct induction followed by resonant magnetic induction, or electromagnetic radiation as microwaves or lasers;

Retinal implant - biomedical implant technology, including epiretinal, subretinal, or suprachoroidal implant, for partially restoring vision to people who have lost it due to degenerative eye conditions such as 'retinitis pigmentosa' or 'macular degeneration';

Holographic televisions - holographic '3D displays' able to depict motion by updating image 30 times a second, for example using a small crystal of lithium niobate;

Cloaking devices - technology successfully tested to make a paperclip invisible;

Hover cars - personal vehicles flying at a constant altitude above water or ground using a cushion of air retained by a flexible skirt and the hovertrain;

Exoskeletons - external skeletons supporting and protecting a living body, built to enhance human capacities;

Force fields - shields of electromagnetic pulses 'EMPs' for protecting military vehicles.

Futurist predictions

Zero-size intelligence in computing for packing a whole lot of brains in a tiny package, and housing for the computer itself to be almost zero;

Moon, Mars and *space* exploration by Space Launch System intending to sent a crew of up to four astronauts into space (NASA);

Neurohacking for reading people's minds with machines, by translating electrical activity from the brain, decoding brainwaves for helping sufferers of dementia or other cerebral disorders, actually highjacking and improving their cerebral activity;

Mass Data for optimizing the choices in buying products and accessing news and information;

Quantum control by improving the atomic-sized power of sound-activated quantum vibrations in computer and communication devices;

Youth Tech movements for gathering the young people's discontent to a organized anarchy or rebellion in form of technological of infrastructure sabotage, either physically or cyberspatially;

Nanotechnology by increasing use not only for innovations in engineering, medical devices, imaging, computing, etc., but also for neurosurgery and gene therapy;

Dark networks, such as 'hacktivism' for computer terror, by incidents of cyber-attacks to water systems or electrical grids;

Universal translators for enabling easier communication across the nations by interconnecting people speaking different languages, so that two differently speaking people to communicate with one another in their own voices but in languages they do not know or understand;

Avatars, *surrogates* and *robotics* will more active roles as replacements for living breathing humans, enabling those with paralysis to move limbs.

Further predictions have been displayed in chronological order as follows:

2015: Windows 9 released by Microsoft; First self-regulating artificial heart; 10 nanometre chips enter mass production;

2016: Holographic versatile disc (HVD) supersedes Blue-ray;

2017: Web-connected video devices exceed the global population; Electronic paper in widespread use;

2018: Many complex surgeries performed by robots; Robot insect spies in military use; Portable, long-range 3D scanning;

2019: Launch of BIOMASS mission; High resolution bionic eyes commercially available;

2020: Internet use reaches 5 billion worldwide; Texting by thinking; Ultra High Definition Television \ Holographic television;

2021: Mind reading technology deployed for security purposes; Traditional microchips reach the limits of miniaturization;

2022: European Extremely Large Telescope operational; Large Synoptic Survey Telescope fully operational;

2023: Brain implants to restore lost memories;

2025: Human brain simulations become possible; Medical robots developed;

2025-35: Advanced Technology Large-Aperture Space Telescope conducting the life-searching mission; Mouse revival from cryopreservation;

2026: Robotic hands matching human capabilities;

2028: Printed electronics become ubiquitous;

2030-39: 'Smart grid' technology widespread;

2031: Web 4.0 transforming Internet landscape;

2033: Holographic wall screens;

2035-40: Swarm robotics reaches nanometre scale;

2037: Quantum computers widely available;

2038: Teleportation of complex organic molecules;

2040: Virtual telepathy dominating personal communications;

2044: Transglobal highway and rail network;

2045: Humans intimately merged with machines;

2047: Fully autonomous, intelligent military aircraft;

2048: Reversible biostasis, organisms' ability to tolerate environmental changes without an active adaptation to them;

2049: Robots become common feature of homes and workplaces.

References

AD14-37: <u>Tiberius Julius Caesar Augustus</u>, 2nd Emperor of Rome, led many campaigns in Europe and Near East, and during his retirement to the island of Capreae (Capri) he initiated a system of communication by *signalling messages with metal mirrors* to reflect the sunlight.

1519-22: <u>Ferdinand Magellan</u>, Portuguese navigator, and <u>Juan Sebastian del Cano</u>, Basque navigator, achieved the first circumnavigation, with five ships including flagship *Santa Maria de la Victoria*, coasting Patagonia, passing through *Magellan Strait*, reaching the *Pacific Ocean* and then *Philippines*, where Magellan was killed by the local people; only one of the ships completed the *first circumnavigation of the world*, being taken back to Spain by the captain JS del Cano.

1608-10: <u>Hans Lippershey</u>, Dutch optician, working in optical devices, discovered that the combination of two separated long-focus and short-focus convex lenses can make distant objects appear nearer, and obtained patent for this type of *telescope*; also he evidenced that if this combination is reversed it becomes microscope.

1632-35: <u>Thomas Withering</u>, English merchant and postal administrator, was commissioned by King Charles I as the Postmaster of Foreign Mails, improved *postal communication* between England and France, and placed the inland mail under a royal monopoly.

1667-68: <u>Isaac Newton</u>, English scientist and mathematician, is credited with constructing the first practical *reflecting telescope*, that was perfected not only for astronomical but also for terrestrial observations related to communication.

1669-1704: <u>Sébastien Le Prestre de Vauban</u>, became famous for his skill in both designing systems of fortification and breaking through them; he wrote *Mémoire pour servir à l'instruction dans la conduite des siège*, and *Instructions pour la défense*; as well as first introduced *ricochet firing* for communication, which was used within his designed three parallel trenches interconnected by zigzagging trenches and tower-bastions along the fortification systems.

1670-78: <u>Christian Huygens</u>, Dutch physicist, stated the *undulatory theory* by the *principle of Huygens*, first proposed the *undulatory theory of light*, discovered the *polarization*, and promoted the *wave theory* as well as the *wave nature of light*.

1735-48: <u>Pierre Bouguer</u>, French physicist, laid the foundation of *photometry*, discovered the *law of absorption* 'Lambert's law', and invented the *heliometer* to measure light of Sun and other luminous bodies.

1746-48: <u>Benjamin Franklin</u>, US statesman, inventor and scientist, made distinction between the *positive electricity* and the *negative electricity*, proved that *lightning* and *electricity* are the same, and suggested that the buildings could be protected by *lightning-conductors*.

1775-87: <u>Alessandro Giuseppe Anastasio Volta</u>, Italian physicist and inventor, discovered the *electrophorus* as precursor of the induction machine, the *condenser*, and the *candle flame collector of atmospheric electricity*, before the invention of electrochemical battery.

1778-82: Richard Howe, English admiral, as commander of the British fleet during the American Revolution, defended the North American coast against a superior French force, and had at his disposal a total of *28 flags* to be seen in conjunction with a *printed code* issued to all his officers.

1785: Charles Augustin Coulomb, French physicist, discovered the law of electrical attraction, called *Coulomb's law*, stating that the force between two small charged spheres is related to charges and distance between them.

1790-95: Luigi Galvani, Italian physiologist, discovered the so-called *animal electricity*, followed by use of discoverer's name in words such as 'galvanized' and 'galvanometer'.

1791-94: Claude Chappe, French engineer and inventor, first developed the idea of a *line of hilltop towers*, each bearing a structure with two hinged arms (each pair of arms movable to any of 49 recognizably different positions, seven for each arm), and every tower having two telescopes focused on its neighbour in either direction, between 3 and 6 miles away; then he demonstrated a practical *semaphore system* opening the way to a modern telecommunication; and finally invented the *non-electric telegraph* as a visual system using semaphore and a flag-based alphabet.

1794-1820: Thomas Young, English physicist, physician and Egyptologist, established the *wave theory of light*, and combined the classical wave theory with the theory of colours to explain the *interference phenomenon* produced by ruled gratings, thin plates, and colours of rainbow.

1799: Henry Edles, British printer, published the *Signal-Book for the Ships of War*, Special Collection Publications, Paper 15, Madras (now preserved in the University of Rhode Island, Kingston, USA).

1800: William Nicholson, English physicist and inventor, constructed the first *voltaic pile in England*; and observed that when the ends of leads from battery were immersed in water, bubbles of gas were produced, thus discovering the *electrolysis*.

1806-22: Thomas Johann Seebeck, Estonian-born German physicist, made studies on heating and chemical effects of the colours of solar spectrum, investigated the optical polarization in stressed glass, and discovered the *thermoelectricity*, initially called 'thermomagnetism', which occurs when an electric current is generated through application of heat to a junction of two metals.

1809-23: Samuel Thomas von Sömmerring, German physician, anatomist and inventor, designed an *electric telegraph*; and then developed the first *telegraphic system* in Bavaria (now in the German Museum of Science, Munich).

1810-22: Jean Baptiste Joseph Fourier, French mathematician and physicist, introduced the expansion of functions in trigonometric series, now known as

Fourier series, by which almost any function of real variable can be expressed as a sum of sines and cosines of integral multiples of variable; stated the *Fourier transform* used in operational calculus; discovered the *greenhouse effect*; developed the technique to resolve partial differential equations for describing the heat conduction in solid bodies; and published his works including *Théorie analytique de la chaleur* 'Analytical Theory of Heat' applying it to solve partial differential equations for heat conduction in solid bodies; *Remarques générales sur l'application du principe de l'analyse algébrique aux équations transcendantes*, and *Mémoire d'analyse sur le mouvement de la chaleur dans les fluides*; as well as discovered an important theorem on the *roots of algebraic equations*.

1814-21: Joseph von Fraunhofer, German physicist, developed the *prism spectrometer* to discover dark lines in Sun's spectrum which now bear his name; then invented the *transmission diffraction grating*, and subsequently the *reflection grating*.

1818-21: Augustin Jean Fresnel, French physicist, invented multi-faceted lighthouse *Fresnel lens*, and the special prism called *Fesnel's rhomb* producing circularly polarized light; established the *undulatory theory of light*; as well as published brilliant papers relating the *polarization phenomena* to the *hypothesis of transverse waves*, and three-volume *Œuvres Complètes* 'Complete Works'.

1820-25: Hans Christian Oersted, Danish physicist, discovered the *magnetic effect produced by an electric current*, paving the way for electromagnetic discoveries of AM Ampère and M Faraday; made extremely accurate measurement of compressibility of water, and first isolated *aluminium*.

1821-43: Johann Carl Friedrich Gauss, German mathematician, astronomer and physicist, developed an early type of *heliograph*, called 'heliotrope' that was used as a marker for geodesic survey and later as a mean of telegraphic communication; studied the theory of *errors of observations*; and gave a mathematical theory of the *optical systems of lenses*.

1822-30: André Marie Ampère, French mathematician and physicist, founded the *science of electrodynamics*, and published *Observations électro-dynamiques* 'Electrodynamic Observations', and *Théories des phénomènes électro-dynamiques* 'Theory of Electrodynamic Phenomena'.

1824: Sadi Carnot, French physicist, published his masterpiece *Réflexions sur la puissance motrice du feu et sur les machines propres à developer cette puissance* 'Reflections on the Motive Power of Fire, and on the Machines Appropriate to develop this Power', concerning on scientific principles to an analysis of *working cycle* and *efficiency of steam engine*.

1824-36: William Sturgeon, British scientist, produced the first *permanent photographs*, making a stable image by a pewter plate coated with bitumen of Judea, a kind of asphalt that hardens on exposure to light; made experiments resulting in invention of the *electromagnet*, as a horseshoe-shaped piece of iron wrapped with 18 turns of bare copper wire, working when a current was passed through the coil, becoming magnetized and attracting other pieces of iron; this invention was followed by the first *moving-coil galvanometer* and various electromagnetic machines described in his *Annals of Electricity*.

1826-28: Harrisson Gray Dyar, US chemist, invented and experimented a frictional *electrolytic telegraph* for sending messages by sparks of electric current which were recorded through chemically treated paper tape to burn dots and dashes, the ribbon being revolved mechanically by a clock apparatus; his telegraph line was used at a race track in Long Island for transmitting, over a half-mile of bare electrical wire, the first message ever sent.

1827: Georg Simon Ohm, German physicist, formulated and published the law relating voltage, current and resistance in an electrical circuit, called *Ohm's law* of charge transport.

1828-30: Jean Baptiste Biot, French physicist and astronomer, and Félix Savart, French physician and physicist, discovered the *law defining the intensity of magnetic field* produced at a given point *near a long straight current-carrying conductor*.

1829-35: Joseph Henry, US physicist, constructed the first *electromagnetic motor*; discovered the electromagnetic phenomenon of *self-inductance*; as well as invented a precursor to the electric *doorbell*, and *electric relay*.

1831-34: Michael Faraday, English chemist and physicist, stated the *law of induction*, showing that the electromotive force produced around a closed path is proportional to the rate of charge of magnetic flux through any surface bounded by that path; founded the *electric motor technology*; as well as defined the *laws of electrolysis*: 1st, mass of substance altered at an electrode is directly proportional to quantity of electricity transferred at that electrode; and 2nd, for a given quantity of electricity, mass of an elemental material altered at an electrode is directly proportional to element's equivalent weight.

1831-44: Samuel Finley Breese Morse, US painter and inventor, experimented with the *single-wire telegraph system*, improved the *electric telegraph*, built the first *long-distance electric telephone line*; and together with Alfred Lewis Vail, US machinist and inventor, developed the *Morse code* signalling alphabet.

1835-37: Antoine César Becquerel, French physicist, investigated the electrical properties of minerals, and first used the *electrolysis to isolate metals from their ores*.

1840-45: James Prescott Joule, English natural philosopher, experimenting on heat, discovered *Joule effect* asserting that heat produced in a wire by an electric current was proportional to the resistance and to the square of the current; showed experimentally that heat is a form of energy; determined quantitatively the amount of mechanical (and later electrical) energy to be expended in the propagation of heat energy, establishing the *mechanical equivalent of heat*; also made a mathematical study of the *current through a resistance causing localized heating*, and first described the phenomenon of *magnetostriction*.

1842: Christian Johann Doppler, Austrian physicist, formulated the *Doppler's principle*, which explains frequency variation observed when a vibrating source of waves and an observer approach or recede from one another, and applies to all forms of electromagnetic radiation, being useful in astronomy, where changing wavelengths of receding celestial bodies provide evidence for expanding universe.

1843-45: Carl Friedrich Gauss and Wilhelm Eduard Weber, German scientists, studied the Earth's magnetism, developed the *magnetometer*, invented the *electrodynamometer*, and first applied the *mirror and scale method* of reading deflections.

1845-49: Armand Hippolyte Louis Fizeau, French physicist, demonstrated the use of shift in light frequency, called *red-shift*, in determining star's speed, which corresponds to CJ Doppler's principle, and should be applied to any wave motion, particularly that of light.

1847-54: George Boole, English mathematician and logician, considering the operations of logic algebraically, he founded the *modern symbolic logic*; and published *Mathematical Analysis of Logic*, and *Laws of Thought*; his well-known *Boolean algebra*, a generalization of the familiar operations of arithmetic, became later the basis in the design of circuits and computers.

1847-67: Ernst Werner von Siemens, German electrical engineer, established factories for making *telegraphy equipment*, a business known as 'Siemens Brothers'.

1848-79: Joseph Wilson Swan, English chemist, experimented with *carbonized paper filaments for electric lamps*, successfully demonstrated a *light bulb*, and patented the *carbon process for printing photographs* in permanent pigment.

1850-55: Alexander Parkes, English chemist and inventor, significantly contributed to the use of *electroplating*, and applied it in the case of a spider's web; also he patented the *xylonite*, a form of celluloid.

1855-80: Adolph Eugen Fick, German physiologist, stated the *laws of diffusion*: 1st, mass of solute diffusing through unit area per second is proportional to concentration gradient; and 2nd, partial derivative of concentration with respect to time is proportional to second derivative of concentration with respect to distance.

1859: Gustav Robert Kirchhoff, German physicist, and Robert Wilhelm Bunsen, German chemist and physicist, discovered the *spectrum analysis*, which facilitated the identification of new elements, including *caesium* and *rubidium*.

1861: Thomas Sutton, English photographer and inventor, created the first *single lens reflex camera*, taking three separate black-and-white photographs of a multicoloured ribbon, one through a blue filter, one through a green filter, and one through a red filter; then he used three projectors equipped with similar filters, the three photographs being projected superimposed on a screen; his ribbon image became known as the *first colour photograph*.

1861-73: James Clerk Maxwell, Scottish physicist, demonstrated *colour photography* with a picture of tartan ribbon; founded the *theory of electromagnetism*, as presented in his published work *Treatise on Electricity and Magnetism*, where M Faraday's theory of electrical and magnetic forces was mathematically stated; suggested that electromagnetic waves could be generated in a laboratory; and paved the way for theory of relativity and quantum mechanics.

1863-78: <u>Ludwig Valentin Lorenz</u>, Danish physicist, and <u>Hendrik Antoon Lorentz</u>, Dutch physicist, worked on *mathematical description for light waves* and on *electron theory* respectively; and independently deduced the *Lorentz-Lorenz formula* relating the *refractive index* of a substance to its *polarizability*.

1865: <u>Rudolf Julius Emmanuel Clausius</u>, German physicist, based on the second law of thermodynamics in relation with the dissipation of energy, he introduced the term *entropy* in such a way that dissipation would be equivalent to increasing entropy.

1865-90: <u>Ernst Abbe</u>, German physicist, developed instruments for measuring refractive indices of glass, and a focometer to control performance of optical workshop; invented the arrangement known as *Abbe's homogeneous immersion*, that was followed by his work on microscope; and founded the *diffraction theory of optical imaging*.

1868-77: <u>Ludwig Boltzmann</u>, Austrian physicist, extending JC Maxwell's theory of velocity distribution for colliding gas molecules, as a basis of *Maxwell-Boltzmann distribution* and *statistics*, he deduced the famous *Boltzmann equation*, a fundamental diffusion equation based on particle conservation, showing how increasing *entropy* corresponds to increasing molecular randomness; and also theoretically stated the law for *black-body radiation*.

1869-71: <u>Zénobe Théophile Gramme</u>, Belgian electrical engineer, constructed the *first successful direct-current dynamo*, incorporating ring-wound armature called *Gramme ring*, and improved it as the *first electric generator* to be used commercially for electroplating and electric lighting.

1870-73: <u>Hermann Wilhelm Vogel</u>, German chemist, invented the *orthochromatic photographic plate*, and made experiments with spectroscopic photography, designing a *photometer*.

1870-80: <u>Wilhelm Eduard Weber</u> and <u>Friedrich Wilhelm Georg Kohlrausch</u>, German physicists, defined the *ratio for charge of a capacitor* in electric and magnetic units, and investigated the *conductivity of electrolytic solutions*, leading to *Kohlrausch's law* on independent ion migration.

1875-77: <u>Alexander Graham Bell</u>, Scottish-born US scientist, inventor and engineer, after experimenting with various acoustical devices, he invented the *telephone* and produced the first intelligible *telephonic transmission*; then patented his telephone, and formed the Bell Telephone Company.

1876-86: <u>Eugene Goldstein</u>, German physicist, showed that *cathode rays* could cast sharp shadows and were emitted perpendicular to the cathode surface; and then published his discovery of *Kanalstrahlen* 'canal rays, or positive rays' emerging from channels or holes in anodes, which later led to the construction of *cathode rays tubes* used in mass spectroscopy, television and computer monitors.

1877-91: <u>Thomas Alva Edison</u>, US inventor and businessman, among numerous invented devices which made him a celebrity, there are the *phonograph*, also called *gramophone*, that first recorded on tinfoil around a grooved cylinder; *incandescent light bulb*; *carbon granule microphone*; a *system for generating and distributing electricity*, as well as *megaphone*, *electric valve* and *kinetoscope*.

1880-92: <u>Elihu Thomson</u>, US inventor, pioneered the electrical manufacturing industry in USA, co-operated in about 700 patented electrical inventions, including *three-phase alternating-current generator* and *arc lighting*; he also founded the Thomson-Houston Electric Company, and then the General Electric Company.

1884-1900: <u>George Eastman</u>, US inventor and philanthropist, produced the successful *rollfilm*, *Kodak box camera*; and together with <u>Thomas Alva Edison</u>, US inventor and physicist, experimented for *moving-picture* industry; then he founded the Eastman Kodak Company, and developed the *Brownie camera* that initially was a cardboard box camera with a single meniscus lens taking pictures on rollfilm.

1885: <u>William Stanley</u>, US electrical engineer, invented the *electric transformer*, and a *long-range transmission system* for alternating current.

1885-88: <u>Nikola Tesla</u>, Croatian-born US physicist and electrical engineer, improved the dynamos and electric motors, and invented the *high-frequency Tesla coil* and the *air-core transformer*.

1887: <u>Heinrich Rudolf Hertz</u>, German physicist, made the fundamental discovery of *Hertzian waves*, now known as *radio waves*, confirming the existence of electromagnetic waves which behave like light waves.

1889-91: <u>Almon Brown Strowger</u>, US inventor, initially constructed a model of telephone from a round collar box and some straight pins, which was developed by him and patented as A.B. Strowger 'Automatic Telephone Exchange', representing the first *direct dial telephone*.

1895: <u>Wilhelm Konrad von Röntgen</u>, German physicist, discovered the *electromagnetic rays*, initially called 'X-rays' because of their unknown properties, later named *Röntgen rays*.

1895-96: <u>Ernest Rutherford</u>, New Zealand physicist, designed the first *wireless system* that was used for the first successful wireless transmission, when he was on a scholarship at Cambridge, England.

1895-1901: <u>Guglielmo Marconi</u>, Italian inventor and electrical engineer, became renowned for his work on *long-distance radio-transmission*, as well as for his development of *Marconi's law* and *radio telegraph* system; he obtained British patent for 'Improvements in Apparatus for Wireless Telegraphy', and sent the first *transatlantic radio signal* from Cornwall to Newfoundland.

1895-1903: <u>Auguste Marie Louis Nicolas Lumière</u> and <u>Louis Jean Lumière</u>,

French industrial physiological chemists and pioneers of motion photography, invented the *cinématographe* 'cinematograph', invented the *autochrome screen plate* for colour photography; produced the first film newsreels and *first motion picture* in history, called *La Sortie des ouvriers de l'usine Lumière* 'Workers Leaving the Lumière Factory'; directed the production of about 2,000 films, including dramatizations of *Faust*, and *The Life and Passion of Jesus Christ*; and finally invented the *autochrome screen plate* for colour photography.

1896: Pieter Zeeman, Dutch physicist, discovered the *Zeeman effect*, showing that the resultant broadening of *spectral emission lines* is due to the splitting of spectrum lines into two or three components.

1897: Ferdinand Braun, German physicist, constructed the first *cathode-ray oscilloscope* 'Braun tube' providing a basic component of television.

1897-1911: Joseph John Thomson, English physicist, and Wilhelm Carl Werner Wien, German physicist and inventor, made similar experiments with the cathode rays, also called *positive rays*, and adding some extra electrodes, they discovered that the deflection of these rays by electric or magnetic fields into glass *tubes* can produce *optic effects*, later used to create the image in a classic television set and also the computer monitors.

1898-1900: Max Karl Ernst Planck, German theoretical physicist, deeply researching the thermodynamics and black body radiation, he abandoned the classical dynamical principles and formulated the *quantum theory*, assuming energy changes to take place in small discrete instalments or *quanta*, predicting phenomena inexplicable in classical Newtonian theory, and resulting in statement of *Planck's radiation law*.

1904: John Ambrose Fleming, English physicist and electrical engineer, invented the thermionic rectifier or *diode valve* 'Fleming valve', which for half a century was a vital part of radio, television, and early computer circuitry.

1906: Reginald Aubrey Fessenden, US radio engineer and inventor, developed the *amplitude modulation*, used for the first *radio broadcast*, and discovered the *heterodyne effect*, which soon was developed into *superheterodyne circuit* as an integral part in design of radio receivers.

1923-28: Louis-Victor de Broglie, French physicist, formulated the *wave-particle duality*, showing that particles can behave as waves, and thus opening the way to wave mechanics.

1924-26: John Logie Baird, Scottish engineer and innovator, invented the world's *first television*, first publicly demonstrated *colour television* system, and first purely electronic *colour television* picture tube.

1926-30: Erwin Schrödinger, Austrian physicist, inspired by LV Broglie's wave-particle duality, he set the basis of the science of *wave mechanics*, as part of quantum theory with celebrated *Schrödinger's wave equation*.

1928-30: Paul Adrien Maurice Dirac, English mathematician and physicist, formulated the *relativistic wave equation*; was author of the classic work *The Principles of Quantum Mechanics*; and introduced the *Dirac delta function*, or unit impulse function (that is zero everywhere except at the single point of origin), that is regarded as a kind of weak limit in a sequence of functions having a tall spike at the signal origin, and making it extremely useful in many natural processes involved in the high technology communication.

1928-32: Harry Nyquist, Swedish electronic engineer, was an important contributor to *communication theory*, and wrote papers such as *Thermal Agitation of Electric Charge in Conductors*, Physical Review, 32; *Certain topics in telegraph transmission theory*, Transactions American Institute of Electrical Engineers, 47; and the classic work on *stability of feedback amplifiers*, called *Regeneration theory*, Bell System Technical Journal, 11.

1928-36: Ralph Vinton Lyon Hartley, US electronics engineer, made researches on the *repeaters* and voice, and carrier transmission; formulated a law showing that the amount of transmissible information is proportional to frequency range and time of transmission; invented *Hartley oscillator* and *Hartley transform*; and published his papers including *Transmission of information*, and *Oscillations in Systems with Non-Linear Reactance*, Bell System Technical Journal, 7(3), and 15(3) respectively.

1928-39: Vladimir Kosma Zworykin, Russian-born US physicist, obtained a patent for applying cathode-ray tube to television, developed the *first practical television camera*, and the *electron microscope*.

1930-55: Ernst Frederick Werner Alexanderson, Swedish-born US electrical engineer and inventor, developed the *Alexanderson alternator* for transoceanic communication, *antenna structures*, and *radio-receiving* and *transmitting systems*, and also a complete *television system* and a *colour television receiver*.

1932-48: Norbert Wiener, US mathematician, studying the stochastic processes and harmonic analysis, he introduced the concepts later called *Wiener integral* and *Wiener measure*; and founded the *cybernetics* that was described in his published work *Cybernetics, or control and communication in the animal and the machine*.

1932-58: Douglas Rayner Hartree, English mathematician and physicist, studied computational methods applied to a wide variety of problems ranging from atomic physics, resulting in invention of the *method of self-consistent field in quantum mechanics*, to the automated control of chemical plants; and developed the *differential analyser*, an *analogue computer*, used as an 'electronic digital computer'.

1937-38: Chester Floyd Carlson, US inventor, made an experiment with copying process using photoconductivity, and discovered the basic principles of *electrostatic xerography*, a non-chemical photographic process in which light discharges a charged dielectric surface.

1941-42: Hedy Lamarr, Austrian actress and inventor, and George Antheil, US avant-garde composer and inventor, were authors of an early technique for *spread spectrum* communication and *frequency hopping*, and patented their 'Secret Communications System' using a code stored on a punched paper tape to synchronize random frequencies, later widely applied to telecommunications.

1941-50: <u>Isaac Asimov</u>, US novelist, critic and popular scientist, first used in print the word *robotics* in his science fiction short story *Liar!*, Astounding Science Fiction; and also wrote other short stories forming the collection *I, Robot*; by these works formulating his *Three Laws of Robotics*.

1942: <u>John Vincent Atanasoff</u> and <u>Clifford Berry</u>, US physicists and computer pioneers, constructed the *electronic calculating machine* 'ABC' (Atanasoff-Berry-Computer), an *early computer* using vacuum tubes.

1945-49: <u>Claude Elwood Shannon</u>, US mathematician, electronic engineer and known as 'the father of information theory', formalized RVL Hartley's storage principle, deducing that the information storable in a system is proportional to the logarithm of number of possible states of the system; on the basis of H Nyquist's paper 'Certain topics in telegraph transmission theory' (1928), he expressed and proved the *Nyquist-Shannon sampling theorem*, as a fundamental bridge between continuous signals and discrete signals, which is applicable to a class of mathematical functions whose Fourier transforms are zero outside of a finite region of frequencies, being provided by the *discrete-time Fourier transform*, and used in *digital signal processing*; thus founding *information theory*, as published in his paper *Communication Theory of Secrecy Systems*, Bell System Technical Journal, 28(4).

1945-53: <u>John von Neumann</u>, Hungarian-born US mathematician and physicist, developed the first *electronic computer*, called *IAS* (Institute for Advanced Study) *machine*, also called 'Neumann machine'; designed a *self-reproducing computer program*, as the world first 'computer virus'; and introduced the *stochastic computing*.

1947: <u>Walter Houser Brattain</u>, <u>John Bardeen</u> and <u>William Bradford Shockley</u>, US physicists, working at Bell Telelephone Laboratories, researched for producing semiconductor devices to replace thermionic valves, and invented the *point-contact transistor*, using a thin germanium crystal. \ <u>Dennis Gabor</u>, Hungarian-born British physicist, conceived the technique of *holography*, as a method of photographically recording and reproducing *three-dimensional images*.

1947-48: <u>Douglas Ring</u> and <u>William Rae Young</u>, US engineers of Bell Labs, invented the *cell phone*, and obtained patent for 'Guided Wave Frequency Range'.

1949: <u>Roberto Busa</u>, Italian Jesuit priest and pioneer of electronic writing, produced his first *e-book* called *Index Thomisticus*, a complete lemmatization of St Thomas Aquinas' works. \ <u>Ángela Ruiz Robles</u>, Spanish teacher, writer and inventor, intending to decrease the number of books that her pupils carried to the school, she patented her first electronic book entitled *La Enciclopedia Mecánica* 'Mechanical Encyclopaedia'.

1951-1953: <u>Charles Hard Townes</u>, US physicist, passing a weak beam of microwaves through excited ammonia gas triggering ammonia molecules to emit their own intense and coherent microwave radiation, he constructed the first operational *maser* '*m*icrowave *a*mplification by *s*timulated *e*mission of *r*adiation'.

1952-65: <u>Alfred Kastler</u>, French scientist, worked to obtain precise information about the *atomic structures* by probing energy levels within atoms, using visible light and radio waves to excite electrons in atoms, which then emitted radiation as they returned to lower energy states; he also used optical techniques to develop *optical pumping*, which laid the foundations for subsequent development of masers and lasers.

1953: <u>Charles Hard Townes</u> and <u>James Gordon</u>, US physicists, built the first *ammonia maser*, a device using stimulated emission in a stream of energized ammonia molecules to produce amplification of microwaves at a frequency of about 24 gigahertz.

1954-89: <u>CERN</u> *(European Organization for Nuclear Research)* was created and the Laboratory for Particle Physics was located just outside of Geneva, Switzerland, with 26.67 kilometres circumference circular underground tunnel that houses the *Large Electron-Positron* (LEP) *collider*, the world's largest and highest-energy synchrotron, which became operational and should produce proton-antiproton collisions in the energy range of 10 - 14 tera-electron-volts.

1955: <u>Nikolai Gennadiyevich Basov</u> and <u>Aleksandr Mikhailovich Prokhorov</u>, Russian physicists, created the first working *laser* '*l*ight *a*mplification by *s*timulated *e*mission of *r*adiation', applicable in spectroscopy, surgical work, *compact disc players*, etc.

1958-70: <u>Clive Marles Sinclair</u>, English electronic engineer and inventor, developed a wide range of *calculators*, *miniature television sets* and *personal computers*, and then manufactured a small three-wheeled 'personal transport' vehicle powered by a washing-machine motor and rechargeable batteries.

1960: <u>Manfred Clynes</u>, US scientist and inventor, and <u>Nathan Kline</u>, US psycho-pharmacologist, published their work *Cyborgs and Space*, Astronautics, where they coined the term of *cyborg* '*cyb*ernetic *org*anism', referring to beings with both biological and artificial parts, who could survive in extraterrestrial environments.

1962-63: <u>Harold Rosen</u>, US engineer at Hughes Aircraft Company, invented the first *operational geosynchronous satellite*, called *Syncom 2*, that was launched on a *Delta rocket* B booster from Cape Canaveral, and used for the world's first *satellite-relayed telephone call*.

1962-64: <u>Pier Georgio Perotto</u>, Italian electrical engineer, after the invention of the *magnetic card system*, he developed *Programma 101* as the first commercial 'desktop personal computer', which was produced by the Italian company Olivetti, and then launched for selling.

1963-64: <u>Louiz Pouzin</u>, French computer scientist, invented the *datagram* '*data* tele*gram*', and participated at the design of *CTSS* 'Compatible *T*ime *S*haring *S*ystem' writing a program called *RUNCOM* '*RUN COM*mands', the first operating *shell* system, i.e. a higher level software interface that permitted execution of contained commands within a folder, as the ancestor of the command-line interface and *shell scripts*.

1963-81: Charles Kuen Kao, Chinese-born Hong Kong British and US electrical engineer and physicist, was appointed head of the electro-optics research group at Standard Telecommunication Laboratories (STE) in Harlow, England, where he pioneered *fiber optics* as a telecommunication medium, demonstrating that the high-loss of existing fibre optics arouse from impurities in the glass, and the silica-based *optical waveguides* can offer a practical way to transmit light via *total internal reflection*; then continued his experiments for application of fiber optics to telecommunication, and published the book *Optical Fiber Technology*, Institute of Electrical & Electronics Engineers (IEEE) Press, New York.

1965: Daniel Stephen Halacy, US writer of non-fiction studies, published his book *Cyborg: Evolution of the Superman*, New York: Harper and Row, featuring an introduction which spoke of a 'new frontier' that was 'not merely space, but more profoundly the relationship between 'inner space' to 'outer space' - a bridge between human and computerized technology, and between 'mind and matter'.

1965-68: Andries van Dam, Dutch-born US computer scientist, worked on electronic devices, coined the term 'electronic book', started the *Hypertext Editing System* (HES) and then developed the *File Retrieval and Editing SyStem* (FRESS) projects, producing *electronic textbooks* for poetry and biology.

1967-71: Erna Schneider Hoover, US mathematician, in order to avoid inconveniences during peak callings on telephone lines, she invented a *computerized telephone switching* method, which 'revolutionized modern communication', preventing system overloads by monitoring call centre traffic, and prioritizing tasks on *phone switching systems* to enable more robust service during peak calling times; her invention was termed 'Feedback Control Monitor for Stored Program Data Processing System'.

1969: Robert (Bob) Kahn and Vinton (Vint) Cerf, US computer scientists and fathers of the Internet, developed the previous *TCP/IP* '*T*ransmission *C*ontrol *P*rotocol / *I*nternet *P*rotocol', as the basic communication language or protocol of the Internet, creating the *ARPANET* '*A*dvanced *R*esearch *P*roject *A*gency *NET*work' and incorporating some designs from French computer scientist L Pouzin.

1969-71: Leonard Kleinrock and Douglas Carl Engelbart, US engineers and inventors, made important contributions to the field of *computer networking*, and developed the *ARPANET*, the precursor of *Internet*, by interconnecting L Kleinrock's Network Measurement Center at UCLA's School of Engineering and Applied Science, and DC Engelbart's NSL system at SRI International in Menlo Park, California.

1970-89: Douglas Carl Engelbart, US computer scientist, became well-known for his invention of *computer mouse*, and contributions in development of many features of modern computing, including *e-mail*, *groupware*, and *hypermedia*.

1971: <u>Federico Faggin</u>, Italian-US, <u>Ted Hoff</u>, US, and <u>Stanley Mazor</u>, US, computer scientists, designed the world's first *microprocessor*, a computer central processing unit usually on one integrated-circuit chip, based on a very large-scale integration technology, called 'Intel 4004'. \ <u>Michael Stern Hart</u>, US author and pioneer of the electronic book 'e-book', created his first *electronic document* by typing the *USA Declaration of Independence* into a computer, and then launched the *Project Guttenberg* to make electronic copies of more texts, especially books.

1971-73: <u>Raymond Samuel Tomlinson</u>, US programmer, implemented an *email* system on the *ARPANET*, being the first system able to send mail between users on different hosts connected to ARPANET; then he introduced the @ *sign* 'at sign' to separate the user from machine, that was adopted in email addresses ever since.

1975-77: <u>Victor Scheinman</u>, US pioneer in the field of robotics, invented the *Programmable universal manipulation arm* (PUMA), a design to *Unimation* product (released at the Robot Type Work Handling Equipment).

1976-78: <u>Donald Ervin Knuth</u>, US computer scientist and mathematician, when writing his seminal series of books 'The Art of Computer Programming (TAOCP)', Stanford University, he took time out to work on typesetting and created the *TeX* - computer typesetting system, related to *METAFONT* - font definition language and rendering system.

1982-2000: <u>Bill Gates</u>, US computer scientist and businessman, had a licence for computer operating system to *International Business Machines* (IBM), fledgling *personal computer* (PC) industry; this system (MS-DOS) was phenomenally successful, and its updated versions, such as *Windows 2000*, allowed maintenance of Microsoft's PC hegemony, and showed how somebody could became a billionaire and world's most wealthy private individual.

1987-95: <u>Jack St Clair Kilby</u>, US electrical engineer, contributed to the development of information and communication by participation at the invention of an *integrated circuit*.

1989-90: <u>Tim Berners-Lee</u>, British computer scientist, and <u>Robert Cailliau</u>, Belgian informatics engineer and computer scientist, created the *World Wide Web* (www) for which they designed and built the first *web* browser at the nuclear physics laboratory (CERN), as the key technology that popularized the *Internet* around the world.

1990-98: <u>Zhores Ivanovich Alferov</u>, Russian physicist, and <u>Herbert Kroemer</u>, German-born US physicist, developed *semiconductor heterostructures used in high-speed-* and *opto-electronics*, as a basic work on *information* and *communication technology*.

1991-92: <u>Neil Papworth</u>, English software architect, designer and developer, working for Sema Group in Reading, Berkshire, England, became known for developing a *Short Message Service* (SMS), by which he sent the world first *text message* 'Merry Christmas'.

1992-96: Joseph Walther, US professor of telecommunication, created the *social information processing theory*; studied the social and interpersonal dynamics of *computer-mediated communication*; and wrote articles including *Interpersonal Effects in Computer-Mediated Interaction: A Relational Perspective*, and *Computer-Mediated Communication: Impersonal, Interpersonal, and Hyperpersonal Interaction*, published in Communication Research, 19(1) and 23(1) respectively.

1993-97: Carl Malamud, US technologist, author and public domain advocate, pioneered *Internet radio*, and launched 'Internet Talk Radio' that was the first *computer-radio talk show*, each week interviewing a computer expert; the first *Internet concert* was broadcasted by the band 'Severe Tire Damage'; then he founded the *Internet Multicasting Service*, and published works such as *A World's Fair for the Global Village*, MIT Press.

1994: Justin Hall, US journalist and entrepreneur, as a student at Swarthmore College, Pennsylvania, became a pioneer of *blogging* by his web-based diary *Justin's Links from the Underground*.

1996-2008: Willard Sterling Boyle, Canadian physicist, and George Elwood Smith, US scientist, invented an *imaging semiconductor circuit*, called *charge-coupled device* 'CCD' *sensor*.

1999: Dave Winer, US software developer and writer, was author of the *Scripting News*, as one of the older and longer running *weblogs*, and became proprietor of a free *blog* service.

1999-2002: Martine Aliana Rothblatt, David Margolese and Robert Brickman, US entrepreneurs, founded the *Sirius Satellite Radio*, which launched the initial phase of its service in the USA.

1999-2006: Mark Warschauer, US professor of education and informatics, published his works *Electronic literacies: Language, culture and power in online education*, Mahwah, New Jersey: Lawrence Erlbaum Associates, and *Laptops and literacy: learning in the wireless classroom*, New York: Teachers College Press.

2002-03: Reid Garret Hoffman, US entrepreneur, venture capitalist and author, and his colleague Allen Blue, founded the *Linkedin* at Santa Monica, California, and then launched it as a social network used primarily for business connections and job searching.

2003-04: Chris DeWolfe and Thomas Anderson, US entrepreneurs, being eUniverse employees, they founded the *Myspace*, with its first version implemented using *ColdFusion*, by the transition from a file storage service to a *social networking site*.

2003-06: Mark Elliot Zuckerberg, US computer programmer and Internet entrepreneur, after graduating at Harvard University, he wrote a program called 'Facemash'; and then together with his colleagues and friends Eduardo Saverin, Andrew McCollum, Dustin Moskovitz and Chris Hughes, developed 'Thefacebook', originally located at thefacebook.com, at Cambridge, Massachusetts, and thus launched the *Facebook*.

252

2005-06: <u>Steve Chen</u> and <u>Jawed Karim</u>, US Internet entrepreneurs, together with <u>Chad Hurley</u>, US designer, founded the *YouTube*, with its first video entitled *Me at the zoo*; then YouTube has been owned by Google.

2006-07: <u>Evan Clark Williams</u>, US internet entrepreneur, <u>Noah Glass</u>, US software developer, <u>Jack Dorsey</u>, US web developer and businessman, and <u>Christopher Isaac 'Biz' Stone</u>, US software engineer and businessman, created the first *Twitter* prototype that was used as an internal service, launched its full version; and then spun off into its own company.

2006-11: <u>Mihai Patrascu</u>, Romanian-born US computer scientist, working on the fundamental questions about *basic data structures*, became known by his papers published such as: with <u>Erik Demaine</u> *Logarithmic lower bounds in the cell-probe model*; with <u>Mikkel Thorup</u> *Higher lower bounds for near-neighbour and further rich problems*; and his own *Unifying the landscape of cell-probe lower bounds*; all in SIAM Journal on Computing, 35(4), 39(2), and 40(3) respectively.

2007: <u>Artemio Ramirez</u> and <u>Shuangyue Zhang</u>, US communication researchers published their work *The effects of the occurrence and timing of modality switching on relationship development*, Communication Monographs, 74(3).

2012: <u>Fabian Bethge</u>, German computational and social researcher, analyzed the relationships between Internet and society, as well as between innovation and organizational communication, and published his findings entitled *Five theses about the future of innovation communication*, Academia.edu, San Francisco, California.

2014: <u>Alice Park</u>, US science writer, published the article *Researchers Clone Cells From Two Adult Men*, TIME, showing that *new human embryonic stem cells* open the way to generate healthy replacements for diseased or damaged cells in patients.